T0141878

Springer Theses

Recognizing Outstanding Ph.D. Research

Aims and Scope

The series "Springer Theses" brings together a selection of the very best Ph.D. theses from around the world and across the physical sciences. Nominated and endorsed by two recognized specialists, each published volume has been selected for its scientific excellence and the high impact of its contents for the pertinent field of research. For greater accessibility to non-specialists, the published versions include an extended introduction, as well as a foreword by the student's supervisor explaining the special relevance of the work for the field. As a whole, the series will provide a valuable resource both for newcomers to the research fields described, and for other scientists seeking detailed background information on special questions. Finally, it provides an accredited documentation of the valuable contributions made by today's younger generation of scientists.

Theses are accepted into the series by invited nomination only and must fulfill all of the following criteria

- They must be written in good English.
- The topic should fall within the confines of Chemistry, Physics, Earth Sciences, Engineering and related interdisciplinary fields such as Materials, Nanoscience, Chemical Engineering, Complex Systems and Biophysics.
- The work reported in the thesis must represent a significant scientific advance.
- If the thesis includes previously published material, permission to reproduce this must be gained from the respective copyright holder.
- They must have been examined and passed during the 12 months prior to nomination.
- Each thesis should include a foreword by the supervisor outlining the significance of its content.
- The theses should have a clearly defined structure including an introduction accessible to scientists not expert in that particular field.

More information about this series at http://www.springer.com/series/8790

Gerard Higgins

A Single Trapped Rydberg Ion

Doctoral Thesis accepted by
Stockholm University, Stockholm, Sweden

Author
Dr. Gerard Higgins
Department of Physics
AlbaNova University Center
Stockholm University
Stockholm, Sweden

Supervisor
Dr. Markus Hennrich
Department of Physics
AlbaNova University Center
Stockholm University
Stockholm, Sweden

ISSN 2190-5053 ISSN 2190-5061 (electronic)
Springer Theses
ISBN 978-3-030-33772-8 ISBN 978-3-030-33770-4 (eBook)
https://doi.org/10.1007/978-3-030-33770-4

This Springer imprint is published by the registered company Springer Nature Switzerland AG
The registered company address is: Gewerbestrasse 11, 6330 Cham, Switzerland

Supervisor's Foreword

Due to tremendous technical progress, we are nowadays able to observe and manipulate individual quantum systems, like single atoms, with a precision that was unthinkable a few decades ago. These technologies have become so mature that we are no longer mere observers of the quantum mechanical properties, but we have started to build prototypes of novel quantum machines.

Non-physicists often find it difficult to understand the underlying quantum physics, as it runs counter to our intuition of the macroscopic world. For instance, if two separated systems are *entangled* then an action on one of the systems can cause an instantaneous effect on the other distant system. It is the counter-intuitive nature of quantum mechanics that makes quantum technologies so powerful and that opens ample possibilities for applications, such as quantum computers that can solve problems intractable by a classical computer, or quantum communication networks that are unconditionally secure.

A range of physical systems are used as prototype quantum machines. In state-of-the-art experiments around 10–20 quantum particles can be entangled. Much effort is being put into scaling up the systems to hundreds or thousands of entangled particles. Such systems could for instance be used as quantum processors with computational power beyond their classical counterparts.

This thesis by Gerard Higgins is concerned with a novel physical platform: trapped Rydberg ions. This new platform combines the advantages of two leading platforms—trapped atomic ions and highly excited Rydberg atoms.

Systems of several trapped atomic ions have led many breakthroughs in this field. Exquisite control of trapped ion systems have been demonstrated, however the generation of entanglement becomes increasingly slow as the system size is increased, making these systems difficult to scale-up.

Systems of neutral atoms excited to Rydberg states have strong interactions which allow for fast generation of entanglement. This has allowed a system of Rydberg atoms to recently break the record of the largest highly entangled GHZ state, which for many years had been held by trapped ion systems.

In Chap. 1, Gerard discusses trapped ion systems and Rydberg atom systems; he describes their key properties, advantages and achievements. In the final part of this chapter, the novel system of trapped Rydberg ions are introduced, the literature in this new field is reviewed and the approaches followed by the two Rydberg ion experiments, one at the University of Mainz and the other here at Stockholm University, are compared. In Chap. 2, Gerard further motivates the novel platform by discussing relevant atomic properties of Rydberg ions, including their interaction strengths.

To conduct this investigation, Gerard and his colleagues set up a new experimental system in a new laboratory, this is described in Chap. 3.

During his Ph.D., Gerard carried out the first two-photon Rydberg excitation of a trapped ion; this has marked advantages over the single-photon excitation which was until recently used at the University of Mainz. The two-photon excitation scheme is presented in Chap. 4.

Unwanted loss of Rydberg ions was the most frustrating part of this work. In Chap. 5, Gerard presents results which show the loss product to be a doubly charged atomic ion.

In the final two chapters, Gerard presents the key results of his thesis:

Chapter 6 deals with investigations of the effects of the trapping electric fields on the sensitive Rydberg ions. Section 6.1 deals with phenomena related to the Rydberg state polarisability, including the altered trapping frequencies experienced by Rydberg ions. Various theory proposals described in Sect. 1.3 utilize this effect. The phenomena in Sect. 6.2 are related to the Rydberg state electric quadrupole moment.

In Chap. 7, Gerard shows Rydberg ions may be coherently excited and deexcited. Rabi oscillations between a low-lying electronic state and a Rydberg state are presented, as well as population transfer using stimulated Raman adiabatic passage (STIRAP).

Gerard's work demonstrates the fundamentals of Rydberg ions as a new quantum technology platform. He demonstrates tiny quantum effects like modified trapping potentials as well as strongly magnified state-dependent forces acting on the Rydberg ions. He shows that Rydberg ions can be controlled with highest precision as required as in quantum technologies. This work has since enabled a sub-microsecond Rydberg ion gate to be conducted, which demonstrates the great potential of this system for the future.

Stockholm, Sweden
August 2019

Dr. Markus Hennrich

Abstract

Systems of trapped ions and systems of ultracold Rydberg atoms are used at the forefront of quantum physics research and they make strong contenders as platforms for quantum technologies. Trapped Rydberg ions are a new hybrid technology envisaged to have both the exquisite control of trapped ion systems and the strong interactions of Rydberg atoms.

In this work a single trapped Rydberg ion is experimentally investigated. A trapped $^{88}Sr^{+}$ ion is excited to Rydberg states using two ultraviolet lasers. Effects of the strong trapping electric fields on the sensitive Rydberg ion are studied. After mitigating unwanted trap effects, the ion is coherently excited to Rydberg states and a quantum gate is demonstrated. This thesis lays much of the experimental groundwork for research using this novel system.

Publications Related to This Thesis

Higgins, G., Li, W., Pokorny, F., Zhang, C., Kress, F., Maier, C., Haag, J., Bodart, Q., Lesanovsky, I. & Hennrich, M. Single Strontium Rydberg Ion Confined in a Paul Trap. *Phys. Rev. X* **7**, 021038 (Jun 2017). http://dx.doi.org/10.1103/PhysRevX.7.021038

Higgins, G., Pokorny, F., Zhang, C., Bodart, Q. & Hennrich, M. Coherent Control of a Single Trapped Rydberg Ion. *Phys. Rev. Lett.* **119**, 220501 (Nov 2017). http://dx.doi.org/10.1103/PhysRevLett.119.220501

Higgins, G., Pokorny, F., Zhang, C. & Hennrich, M. Highly-polarizable Rydberg ion in a Paul trap. *Phys. Rev. Lett.* **123**, 153602 (Oct 2019). http://dx.doi.org/10.1103/PhysRevLett.123.153602

Acknowledgements

First and foremost I thank my supervisor, Markus Hennrich. I count myself lucky having a supervisor who is always aware of what is going on in the lab and who provides constant, patient guidance.

From the time when the lab was empty, it has been a pleasure to work with Fabian Pokorny. Thank you for your steadfast good cheer and unwavering calm. I also thank Christine, Hannes, Florian, Quentin, Chi and Andreas for many good times together in the lab.

During my time in Innsbruck, I was fortunate to be part of the Blatt group, in which expertise is generously shared. In particular, I wish to thank Matthias Brandl for advice about electronics, Muir Kumph for guidance in building stable optical resonators, Esteban Martinez for answering countless questions about trapped ion experiments, Michael Niedermayr for gold-coating the trap electrodes, Ben Ames for help crystallising ions, the senior scientists Christian Roos, Tracy Northup and Yves Colombe for sharing their experience; and Rainer Blatt for welcoming me into his group. I also thank the members of the electronic and mechanical workshops for much assistance.

I wish to thank the administrative staff, both in Innsbruck and in Stockholm, for help with bureaucratic issues; particular thanks are due to Patricia Moser and Isa Callderyd Öhman. I would also like to thank Sandra Scherl, Per-Erik Tegnér and Georg Moser for help arranging the cotutelle agreement.

I thank the theorists Weibin Li and Igor Lesanovsky at the University of Nottingham for a fruitful collaboration.

I thank Nadine, Gabriel, Esteban, Martin, Slava, Philip, Bojana and Bárbara in Innsbruck and Massi, Irina, Vani, Colin, Belu, Tania and Iara in Stockholm for friendship.

Thank you Cris for support and for so many adventures; I love you to bits.

Lastly I thank my mother for always encouraging me and for making me feel confident.

Contents

Acronyms

AOM	Acousto-optic modulator
DDS	Direct digital synthesizer
EMCCD	Electron-multiplying charge-coupled device
MW	Microwave
PDH	Pound–Drever–Hall
PMT	Photo-multiplier tube
RF	Radiofrequency
SFG	Sum frequency generation
SHG	Second harmonic generation
STIRAP	Stimulated Raman adiabatic passage
UV	Ultraviolet
VUV	Vacuum-ultraviolet

Chapter 1
Introduction

In the last decades researchers have achieved exquisite control of different quantum systems. This has allowed fundamental tests to be carried out, including studies of quantum measurements [1, 2], quantum contextuality [3] and the wave function [4, 5]; as well as Bell test experiments [6–8]. New technologies which take advantage of highly-controlled quantum systems are being pursued [9], and proof of principle devices capable of quantum computation [10, 11], quantum simulation [12], quantum communication [13] and quantum metrology [14] have been demonstrated.

Some systems are more suitable than others as platforms for particular quantum technologies, just as some systems are more suitable than others for carrying out particular fundamental tests. For instance, the long coherence times of trapped ion qubits make for excellent quantum memories [15], while the propagation speed of photonic qubits allows the locality loophole to be closed in Bell test experiments. My thesis is concerned with a new experimental platform, namely a system of trapped Rydberg ions. This platform combines two established systems: trapped atomic ions and Rydberg atoms. In this opening chapter aspects of the two constituent technologies are summarised before trapped Rydberg ions are introduced.

1.1 Trapped Atomic Ions

Atomic ions may be trapped in electromagnetic fields. The trap configuration typically used for quantum information purposes is called a linear Paul trap [16]. In such a trap a string of ions is confined by a combination of oscillating and static electric fields. Quantum bits (qubits) are stored in electronic states of the ions. Qubits are coherently manipulated using lasers (and sometimes microwaves) which drive transitions between electronic states. Ions can be well isolated from the environment; they are trapped in ultra-high vacuum, usually around 30 to 300 μm from any surface [17]. Owing in part to this, trapped ion qubits show excellent coherence properties; coherence times of several minutes have been demonstrated [15]. Ions can be confined for days in the deep potential of a Paul trap [18].

G. Higgins, *A Single Trapped Rydberg Ion*, Springer Theses,
https://doi.org/10.1007/978-3-030-33770-4_1

Ion qubits can be prepared, manipulated and read-out with high fidelity [19]. Entanglement operations between ion qubits can be carried out with low errors [20, 21]. A system of 14 trapped ion qubits holds the record for the largest genuine multipartite entangled state stored on separate particles.[1] The trapped ion architecture is a leading contender for quantum computation and simulations.

Quantum simulations with trapped ions have allowed researchers to study exotic phenomena, such as particle-antiparticle production in a lattice gauge field model [23], a discrete time crystal [24], many-body localisation [25] and a dynamical phase transition involving 53 spins [26]. Systems of trapped ions have been used to simulate open quantum systems [27], to study statistical mechanics in quantum systems [28–30] and to investigate thermodynamics at the level of a single atom [31]. They may be employed in the future for experimental investigations in the emerging field of quantum thermodynamics. Quantum simulators are also of commercial interest; they may be utilised to find molecular energies and this may assist pharmaceutical research [32]. A trapped ion quantum simulator has already been used to find the energies of a simple diatomic molecule [33].

Trapped ions have been used as proof-of-principle quantum computers. Of the various algorithms proposed for quantum computers, Shor's factorisation algorithm has generated the most interest, since it could be used to break public-key cryptography schemes. While condensed versions of this algorithm have been demonstrated on various platforms [34–37], the first scalable demonstration was carried out with trapped ions [11]. Various quantum error correction codes have been demonstrated in systems of trapped ions [38–40], this shows fault-tolerant quantum computing may be feasible if technological hurdles are overcome.

Trapped ions may be employed in the future for quantum communication. The workings of a quantum network in which photons transport quantum information between trapped ion memories has been demonstrated [13].

Another potential application of trapped ions is in quantum metrology; entangled states have been engineered for precision spectroscopy with enhanced sensitivity [41] and with less susceptibility to noise [42]. Further, some of the most accurate atomic clocks are trapped ion systems [43, 44].

Trapped ion quantum computers and simulators have yet to out-perform classical computers. One of the sticking points is that manipulation of the entanglement of trapped ion qubits becomes more difficult as the number of ions in a string is increased. Entanglement manipulation in trapped ion systems commonly involves addressing of individual motional modes [45]. As the number of ions in a string is increased the number of motional modes increases and longer laser pulses are required for individual modes to be frequency-resolved [46]. Entanglement manipulation may also be carried out by coupling electronic degrees of freedom to many motional modes simultaneously, however this strategy also requires longer laser pulses as the number of ions in a string is increased [47]. As the duration of

[1]A recent preprint reports a genuine multipartite entangled state of 18 qubits stored on 6 photons [22].

Table 1.1 Scaling of properties of Rydberg atoms and ions in terms of the principal quantum number n and the core charge (also called effective nuclear charge) Z [50, 51]. $Z = +1$ for neutral Rydberg atoms, $Z = +2$ for Rydberg ions with +e overall charge

Property	n-scaling	Z-scaling
Binding energy E_n	n^{-2}	Z^2
Energy separation $E_{n+1} - E_n$	n^{-3}	Z^2
Fine structure splitting	n^{-3}	Z^4
Orbital size $\langle r \rangle$	n^2	Z^{-1}
Electric quadrupole moment $\Theta \sim \langle r^2 \rangle$	n^4	Z^{-2}
Natural lifetime τ_{nat}	n^3	Z^{-4}
Blackbody radiation limited lifetime τ_{BBR}	n^2	Z^{-4}
Transition dipole moment of Rydberg excitation transition $\langle g\|er\|nLJ \rangle$	$n^{-3/2}$	Z^{-1}
Transition dipole moment $\langle nL'J'\|er\|nLJ \rangle$	n^2	Z^{-1}
Dipole-dipole interaction strength	n^4	Z^{-2}
Electric polarisability ρ	n^7	Z^{-4}
Van der Waals coefficient C_6	n^{11}	Z^{-6}

quantum operations increases, errors from decoherence become significant and quantum operations become unfeasible.

To scale up trapped ion quantum systems to the point where they out-perform classical computers, several groups are working towards a ‘quantum charge-coupled device’ architecture in which ions are shuttled between interconnected traps and the entanglement of small numbers of ion qubits are manipulated at a time [48]. Such an architecture likely involves a two-dimensional trapping geometry and may be better suited for simulation of a two-dimensional quantum system than the one-dimensional trapping geometry of a single linear Paul trap [12].

An alternate path to a scalable system was proposed by Müller et al. [49]. They suggest combining trapped ion and Rydberg atom systems to give a novel platform, the experimental investigation of which is the topic of this thesis. Neutral Rydberg atoms are introduced in the next section and the proposal from Müller et al. is described in Sect. 1.3.1.

1.2 Rydberg Atoms

Atomic states with high principal quantum numbers $n \gg 1$ are called Rydberg states. Rydberg atoms have exaggerated properties which follow scaling relations. Some of these properties are presented in Table 1.1.

Much of the current research with Rydberg atoms employs strong, long-range, dipole-dipole interactions between them [52]. Depending on the separation between Rydberg atoms and the relation between the Rydberg states employed, dipole-dipole interactions are manifest as van der Waals interactions, Förster interactions or resonant dipole-dipole interactions. Individual trapped atoms excited to strongly-interacting Rydberg states have been used for the first deterministic generation of entanglement between two neutral atoms [53], for two-qubit Rydberg gates [54] and for experimental investigations of many-body spin models [55]. Atoms have been individually trapped and excited to Rydberg states from arrays of optical tweezers, arrays of magnetic traps and from optical lattices [56]. Two-dimensional trapping configurations are often used, which, in terms of quantum computing and simulation, has scalability advantages over the one-dimensional configuration typical of trapped ion systems.

In a Rydberg atom quantum computer qubits are encoded in Zeeman sublevels or hyperfine sublevels of the ground states of atoms. Single-qubit rotations are carried out using microwaves (MWs) or by two-photon Raman transitions, and two-qubit manipulations are carried out using the strong interactions between atoms excited to Rydberg states.

Neutral Rydberg atom quantum computers and simulators have technological hurdles that need to be overcome before they can compete with trapped ion quantum computers and simulators, let alone out-perform classical computers. A major obstacle in these systems is loss of ground state atoms from the trapping potential and atom loss during Rydberg excitation; these challenges are reviewed in [56]. Atom loss during Rydberg excitation also poses a problem in trapped Rydberg ion systems, as is discussed in Chap. 5.

Optical nonlinearities are generally too weak for nonlinear optical phenomena, such as photon-photon gates, to be observed at the few-photon level. Strong interactions between Rydberg atoms mean that strong optical nonlinearities can be experienced by just a few photons resonant with a transition to a Rydberg state in an atomic cloud [57].[2]

Strong Rydberg interactions give rise to exotic states of matter; as a result of strong interaction between Rydberg atoms and ground state atoms weakly-bound molecules have been produced with bond lengths of thousands of Bohr radii [60], and with giant electric dipole moments [61].

In terms of metrology, Rydberg atoms have potential as quantum-enhanced sensors of electric fields [14] and as surface probes [62, 63] owing to their high electric polarisabilities. High transition dipole moments between Rydberg states make Rydberg atoms excellent sensors of weak radiofrequency (RF) fields [64].

[2]While Rydberg-mediated nonlinear quantum optics is an exciting research field, it is unlikely to be closely linked to the novel trapped Rydberg ion system explored in this thesis. This is because trapped ion systems have much lower optical depths than atomic clouds. However, cavities could conceivably be used to enhance coupling between ions and either ultraviolet (UV) Rydberg-excitation photons or MW photons which couple Rydberg states [58, 59], to allow for nonlinear quantum optics mediated by Rydberg ions.

1.3 Trapped Rydberg Ions

Compared with Rydberg atoms, Rydberg ions have historically received little attention. The valence electron of a positively-charged ion, with core charge (also called effective nuclear charge) $Z = +2$, is more tightly bound than the valence electron in an atom, which has $Z = +1$. Thus, more energy is required for excitation of Rydberg ions than for Rydberg atoms. Other differences between Rydberg atom and Rydberg ion properties in terms of Z are shown in Table 1.1.

Rydberg ion spectra were first studied systematically in the eighties [65–67]. Atoms from atomic beams were photoionised using pulsed lasers and then excited to Rydberg states via multi-photon transitions. Rydberg ions were then doubly-ionised and detected. These studies were motivated by a desire to provide insights into "planetary atoms" with two Rydberg electrons, which are difficult to model because of electron correlations [67].

The high sensitivity of Rydberg states to electric fields and the prohibitive Rydberg excitation energy requirements did little to encourage experimentalists to excite trapped ions to Rydberg states. This changed after a theoretical investigation by Müller et al. [49] predicted trapping of Rydberg ions is feasible and strong interactions between Rydberg-excited ions may be used for fast quantum gates. This prompted two experiments to strive for trapped Rydberg ions; it has also encouraged further theoretical investigations. The theoretical investigations are reviewed in Sect. 1.3.1 and the experiments are introduced and compared in Sect. 1.3.2.

1.3.1 Theoretical Investigations

In 2008 Müller et al. [49] predicted it is feasible to Rydberg-excite ions confined in a linear Paul trap, despite the high sensitivity of Rydberg states to electric fields (see Table 1.1). Although strong field gradients ($\sim 10^9$ V m^{-2}) are used for trapping, ions are well localised to within tens of nanometres from the electric field null. Müller et al. predicted for Rydberg ions up to at least $n = 50$ the electric field of the ionic core dwarfs the field of the trap and Rydberg electrons are bound to the ionic core.

They did, however, predict two effects of the trap on the Rydberg ion, which are studied in Chap. 6. Firstly, Rydberg ions experience a different trapping potential to low-lying ions due to their large electric polarisabilities, and so the trapping frequencies in Rydberg states are altered. This causes unwanted entanglement to be generated between electronic and motional degrees of freedom of ions during Rydberg excitation. Secondly, states with $J > \frac{1}{2}$ have giant electric quadrupole moments and the strong electric quadrupole fields used for ion trapping strongly couple such Rydberg states. Müller et al. suggest mitigating this second trap effect by using states with $J = \frac{1}{2}$: $nS_{1/2}$ and $nP_{1/2}$.

Müller et al. pointed out that although van der Waals interactions are significantly weaker between Rydberg ions than they are between neutral Rydberg atoms (see

Table 1.1), sizeable interactions between Rydberg ions may be generated by using MW radiation to couple Rydberg $nS_{1/2}$ and $n'P_{1/2}$ states. The MW-dressed states can have large dipole moments which rotate with the frequency of the MW radiation. Two MW-dressed Rydberg ions may then interact strongly via dipole-dipole interaction. Müller et al. proposed a trapped ion quantum computer in which qubits are stored in low-lying electronic states and two-qubit gates are carried out by exciting ion qubits to strongly-interacting Rydberg states, similar to the Rydberg gates implemented in a system of neutral atoms [54].[3] Such Rydberg gates are envisaged to be fast (see Sect. 2.6) and offer an alternate route towards a scalable trapped ion quantum computer or simulator. They also proposed a scheme for simulating a spin model in a chain of trapped Rydberg ions.

To mitigate unwanted entanglement between electronic and motional degrees of freedom of ions which is generated during Rydberg excitation (discussed by Müller et al. [49]), in 2014 Li et al. [68] proposed engineering interacting MW-dressed Rydberg states with vanishing electric polarisabilities. Such dressed states should exist because $nS_{1/2}$ and $n'P_{1/2}$ states have polarisabilities with opposite signs.[4] Unfortunately such dressed states have non-maximal dipole moments and thus they do not produce the strongest interactions.

In 2013 Li et al. [69] proposed shaping the motional mode spectrum of an ion crystal using Rydberg ions. They showed that the motional mode structure of a linear string of 100 ions is significantly altered when two ions are excited to Rydberg states, due to the altered trapping potential experienced by Rydberg ions [49]. Motional modes appear which are localised on the ions that lie between the Rydberg ions. The entanglement between these Rydberg-flanked ions may then be manipulated using the common protocol in which the localised motional modes are used as an intermediary [45]. After deexcitation from the Rydberg state the vibrational mode spectrum returns to normal. In this proposal Rydberg ions are not used directly for qubit manipulation, but rather for segmenting the string. In 2015 Nath et al. [70] suggested using the shaping of the motional mode spectrum by Rydberg ions as a way to implement a quantum magnetism simulation in a two-dimensional ion crystal [70].

In a linear Paul trap ions may form linear strings or zigzag crystals, depending on the ratio of the radial mode frequencies and the axial mode frequency. A transition between these two structures can be induced by changing the radial mode frequencies. In 2012 Li et al. [71] proposed starting with a system close to the structural transition and inducing the structural phase transition by exciting an ion to a Rydberg state, then reversing the transition by deexcitation. The transition is induced because the Rydberg ion experiences altered trapping frequencies [49]. Large coherent forces

[3]Rydberg states are not used for storing qubits because the lifetimes of the low-lying states typically used as qubits ($\sim$1 s) are around five orders of magnitude longer than Rydberg states with $n = 50$ ($\sim$10 μs), while similar Rabi frequencies are used to manipulate low-lying states and to excite Rydberg states ($\sim 2\pi \times 1$ MHz).

[4]Throughout this thesis Russell-Saunders term symbols L_J describe total angular momentum quantum numbers. The multiplicity $2S + 1 = 2$ is omitted.

accompany the transition and a range of potential applications are detailed in the proposal.

The principle behind Rydberg mode shaping and Rydberg-induced structural phase transitions was experimentally demonstrated in [72], though Rydberg ions were not used. Instead $^{40}Ca^{+}$ ions were irreversibly doubly-ionised and a $^{40}Ca^{+}$ ion crystal was compared with a mixed species ion crystal of $^{40}Ca^{+}$ and $^{40}Ca^{2+}$.

1.3.2 Experimental Realisations

Trapped ions are excited to Rydberg states in two experiments: $^{40}Ca^{+}$ ions are used in the group of Ferdinand Schmidt-Kaler at the University of Mainz and $^{88}Sr^{+}$ ions are excited in the group of Markus Hennrich, which started in 2012 at the University of Innsbruck and moved to Stockholm University in 2015. This thesis was carried out in the $^{88}Sr^{+}$ experiment.

Both ion species are alkaline earth metals, with one valence electron. Both have $S_{1/2}$ ground states and metastable $D_{3/2}$ and $D_{5/2}$ states with lifetimes $\sim$1 s. Neither species has a hyperfine structure, and in each species optical qubits are stored in a Zeeman sublevel of the $S_{1/2}$ ground state and a Zeeman sublevel of the metastable $D_{5/2}$ state. In both experiments ions are excited to Rydberg states from either of the D-states.

In the $^{40}Ca^{+}$ experiment ions are excited to Rydberg states by driving a single-photon transition using vacuum-ultraviolet (VUV) laser light between 122 and 123 nm. From the metastable $3D$-states electric dipole transition selection rules allow excitation to either Rydberg P- or Rydberg F-states. Thus far they have excited only Rydberg F-states; they do not know why P-states have been unattainable.

In the $^{88}Sr^{+}$ experiment we excite ions in a two-photon excitation scheme using 243 nm and 304 nm to 309 nm ultraviolet (UV) laser light. Excitation proceeds from the metastable $4D$-states to Rydberg S- or Rydberg D-states using excitation laser light which is detuned from resonance with the intermediate $6P$-states, as shown in Fig. 1.1.

The $^{40}Ca^{+}$ experiment began first and achieved the first trapped Rydberg ions [73, 74]. Further progress in this experiment has been impeded by the broad linewidths of their Rydberg resonance lines ($\approx 2\pi \times 4$ MHz for 22F, $2\pi \times 60$ MHz for 52F [75]), which makes investigation of trap effects difficult and coherent excitation of Rydberg states unobtainable. The linewidths have technical contributions, including the VUV laser linewidth, uncompensated electric dipole fields which perturb the Rydberg state, and the poorly-defined VUV laser polarisation which causes multiple non-degenerate transitions to be driven—these transitions are not individually resolved. The linewidths are also Doppler broadened because the relatively high momentum of a VUV photon makes it difficult to access the Lamb-Dicke regime (Sect. 3.1.1). Additionally the oscillating electric trapping field couples different Rydberg states, this leads to non-resolved Floquet sidebands (Sect. 6.2) which also contribute to the

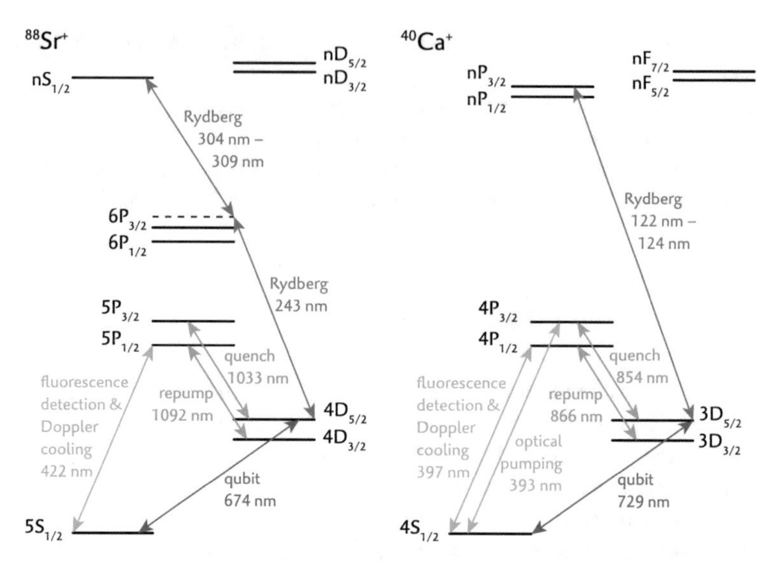

Fig. 1.1 The ion species and laser systems in the two trapped Rydberg ion experiments. In Stockholm S- and D-Rydberg states of $^{88}\text{Sr}^{+}$ are excited via two-photon transitions, while in Mainz P- and F-Rydberg states of $^{40}\text{Ca}^{+}$ may be excited via single-photon transitions. In both systems qubits are stored in sublevels of the ground state and the metastable $D_{5/2}$ state

linewidth of the 52F resonance. A further experimental obstacle is the difficulty of maintaining the VUV Rydberg-excitation laser.

In the $^{88}\text{Sr}^{+}$ experiment we have progressed further. We measure narrower Rydberg resonances ($\approx 2\pi \times 300\,\text{kHz}$, likely limited by the Rydberg-excitation lasers linewidths) thanks to our state-of-the-art ion trap (Sect. 3.2), the counter-propagating two-photon Rydberg excitation setup (Sect. 4.4), addressing of individual transitions (Sect. 4.5) and near-ground-state cooling which allows effects of the trapping electric fields on the highly-polarisable Rydberg ion to be mitigated (Sect. 6.1). The narrow Rydberg resonances allow us to investigate effects of the trap on a Rydberg ion (Chap. 6) and to coherently excite Rydberg states (Chap. 7).

Our experiment proceeds despite an objection raised to the two-photon excitation scheme; higher laser powers are required in a multi-photon excitation scheme, which in turn make ion loss by photoionisation more likely [76]. Ion loss is the topic of Chap. 5.

References

1. Itano WM (2009) Perspectives on the quantum Zeno paradox. J Phys Conf Ser 196:012018
2. Kocsis S et al (2011) Observing the average trajectories of single photons in a two-slit interferometer. Science 332:1170–1173

3. Lapkiewicz R et al (2011) Experimental non-classicality of an indivisible quantum system. Nature 474:490
4. Ringbauer M et al (2015) Measurements on the reality of the wavefunction. Nat Phys 11:249
5. Nigg D et al (2016) Can different quantum state vectors correspond to the same physical state? An experimental test. New J Phys 18:013007
6. Shalm LK et al (2015) Strong loophole-free test of local realism. Phys Rev Lett 115:250402
7. Giustina M et al (2015) Significant-loophole-free test of bell's theorem with entangled photons. Phys Rev Lett 115:250401
8. Hensen B et al (2015) Loophole-free bell inequality violation using electron spins separated by 1.3 kilometres. Nature 526:682–686
9. Quantum Technologies Roadmap (2016) http://qurope.eu/system/files/QT%20Roadmap%202016_0.pdf. Accessed 29 Jan 2018
10. Ladd TD et al (2010) Quantum computers. Nature 464:45
11. Monz T et al (2016) Realization of a scalable Shor algorithm. Science 351:1068–1070
12. Blatt R, Roos CF (2012) Quantum simulations with trapped ions. Nat Phys 8:277
13. Duan L-M, Monroe C (2010) Colloquium: quantum networks with trapped ions. Rev Mod Phys 82:1209–1224
14. Facon A et al (2016) A sensitive electrometer based on a Rydberg atom in a Schrödinger-cat state. Nature 535:262–265
15. Wang Y et al (2017) Single-qubit quantum memory exceeding tenminute coherence time. Nat Photonics 11:646–650
16. Major FG, Gheorghe VN, Werth G (2005) Charged particle traps. Springer, Berlin
17. Brownnutt M, Kumph M, Rabl P, Blatt R (2015) Ion-trap measurements of electric-field noise near surfaces. Rev Mod Phys 87:1419–1482
18. Werth G, Gheorghe VN, Major FG (2009) Charged particle traps II: applications. Springer, Berlin
19. Harty TP et al (2014) High-fidelity preparation, gates, memory, and readout of a trapped-ion quantum bit. Phys Rev Lett 113:220501
20. Ballance CJ, Harty TP, Linke NM, Sepiol MA, Lucas DM (2016) High-fidelity quantum logic gates using trapped-ion hyperfine qubits. Phys Rev Lett 117:060504
21. Gaebler JP et al (2016) High-fidelity universal gate set for 9Be+ ion qubits. Phys Rev Lett 117:060505
22. Wang X-L et al (2018) 18-qubit entanglement with photon's three degrees of freedom. arXiv: 1801.04043
23. Martinez EA et al (2016) Real-time dynamics of lattice gauge theories with a few-qubit quantum computer. Nature 534:516
24. Zhang J et al (2017) Observation of a discrete time crystal. Nature 543:217
25. Smith J et al (2016) Many-body localization in a quantum simulator with programmable random disorder. Nat Phys 12:907
26. Zhang J et al (2017) Observation of a many-body dynamical phase transition with a 53-qubit quantum simulator. Nature 551:601
27. Barreiro JT et al (2011) An open-system quantum simulator with trapped ions. Nature 470:486
28. Clos G, Porras D, Warring U, Schaetz T (2016) Time-resolved observation of thermalization in an isolated quantum system. Phys Rev Lett 117:170401
29. Neyenhuis B et al (2017) Observation of prethermalization in long-range interacting spin chains. Sci Adv 3(8):e1700672
30. Eisert J, Friesdorf M, Gogolin C (2015) Quantum many-body systems out of equilibrium. Nat Phys 11:124
31. Roßnagel J et al (2016) A single-atom heat engine. Science 352:325–329
32. O'Malley PJJ et al (2016) Scalable quantum simulation of molecular energies. Phys Rev X 6:031007
33. Shen Y et al (2017) Quantum implementation of the unitary coupled cluster for simulating molecular electronic structure. Phys Rev A 95:020501

34. Vandersypen LMK et al (2001) Experimental realization of Shor's quantum factoring algorithm using nuclear magnetic resonance. Nature 414:883
35. Lu C-Y, Browne DE, Yang T, Pan J-W (2007) Demonstration of a compiled version of Shor's quantum factoring algorithm using photonic qubits. Phys Rev Lett 99:250504
36. Lanyon BP et al (2007) Experimental demonstration of a compiled version of Shor's algorithm with quantum entanglement. Phys Rev Lett 99:250505
37. Lucero E et al (2012) Computing prime factors with a Josephson phase qubit quantum processor. Nat Phys 8:719
38. Chiaverini J et al (2004) Realization of quantum error correction. Nature 432:602
39. Schindler P et al (2011) Experimental repetitive quantum error correction. Science 332:1059–1061
40. Nigg D et al (2014) Quantum computations on a topologically encoded qubit. Science 345:302–305
41. Leibfried D et al (2004) Toward Heisenberg-limited spectroscopy with multiparticle entangled states. Science 304:1476–1478
42. Roos CF, Chwalla M, Kim K, Riebe M, Blatt R (2006) 'Designer atoms' for quantum metrology. Nature 443:316
43. Chou CW, Hume DB, Koelemeij JCJ, Wineland DJ, Rosenband T (2010) Frequency comparison of two high-accuracy Al+ optical clocks. Phys Rev Lett 104:070802
44. Huntemann N, Sanner C, Lipphardt B, Tamm C, Peik E (2016) Single-ion atomic clock with 3×10^{-18} systematic uncertainty. Phys Rev Lett 116:063001
45. Mølmer K, Sørensen A (1999) Multiparticle entanglement of hot trapped ions. Phys Rev Lett 82:1835–1838
46. Wineland DJ et al (1998) Experimental issues in coherent quantum-state manipulation of trapped atomic ions. J Res Natl Inst Stand Technol 103:259–328
47. Korenblit S et al (2012) Quantum simulation of spin models on an arbitrarylattice with trapped ions. New J Phys 14:095024
48. Kielpinski D, Monroe C, Wineland DJ (2002) Architecture for a large-scale ion-trap quantum computer. Nature 417:709
49. Müller M, Liang L, Lesanovsky I, Zoller P (2008) Trapped Rydberg ions: from spin chains to fast quantum gates. New J Phys 10:093009
50. Lebedev V, Beigman I (1998) Physics of highly excited atoms and ions. Springer, Berlin
51. Bethe HA, Salpeter EE (1977) Quantum mechanics of one- and two-electron atoms. Springer, Boston, MA
52. Browaeys A, Barredo D, Lahaye T (2016) Experimental investigations of dipole-dipole interactions between a few Rydberg atoms. J Phys B 49:152001
53. Hagley E et al (1997) Generation of Einstein-Podolsky-Rosen pairs of atoms. Phys Rev Lett 79:1–5
54. Isenhower L et al (2010) Demonstration of a neutral atom controlled-NOT quantum gate. Phys Rev Lett 104:010503
55. Gross C, Bloch I (2017) Quantum simulations with ultracold atoms in optical lattices. Science 357:995–1001
56. Saffman M (2016) Quantum computing with atomic qubits and Rydberg interactions: progresss s and challenges. J Phys B 49:202001
57. Firstenberg O, Adams CS, Hofferberth S (2016) Nonlinear quantum optics mediated by Rydberg interactions. J Phys B 49:152003
58. Miller R et al (2005) Trapped atoms in cavity QED: coupling quantized light and matter. J Phys B 38:S551
59. Raimond JM, Brune M, Haroche S (2001) Manipulating quantum entanglement with atoms and photons in a cavity. Rev Mod Phys 73:565–582
60. Bendkowsky V et al (2009) Observation of ultralong-range Rydberg molecules. Nature 458:1005
61. Niederprüm T et al (2016) Observation of pendular butterfly Rydberg molecules. Nat Commun 7:12820

62. Kübler H, Shaffer JP, Baluktsian T, Löw R, Pfau T (2010) Coherent excitation of Rydberg atoms in micrometre-sized atomic vapour cells. Nat Photonics 4:112
63. Tauschinsky A, Thijssen RMT, Whitlock S, van Linden van den Heuvell HB, Spreeuw RJC (2010) Spatially resolved excitation of Rydberg atoms and surface effects on an atom chip. Phys Rev A 81:063411
64. Fan H et al (2015) Atom based RF electric field sensing. J Phys B 48:202001
65. Boulmer J, Camus P, Gagne JM, Pillet P (1987) Laser-microwave ionisation mass spectroscopy of ion Rydberg states: Ba+ spectrum. J Phys B 20:L143
66. Jones RR, Gallagher TF (1989) Observation of $Ba^{+}np_{1/2}$ and ngj Rydberg series. J Opt Soc Am B 6:1467–1472
67. Lange V, Khan MA, Eichmann U, Sandner W (1991) Rydberg states of the strontium ion. Z Phys D 18:319–324
68. Li W, Lesanovsky I (2014) Entangling quantum gate in trapped ions via Rydberg blockade. Appl Phys B 114:37–44
69. Li W, Glaetzle AW, Nath R, Lesanovsky I (2013) Parallel execution of quantum gates in a long linear ion chain via Rydberg mode shaping. Phys Rev A 87:052304
70. Nath R et al (2015) Hexagonal plaquette spin-spin interactions and quantum magnetism in a two-dimensional ion crystal. New J Phys 17:065018
71. Li W, Lesanovsky I (2012) Electronically excited cold ion crystals. Phys Rev Lett 108:023003
72. Feldker T et al (2014) Mode shaping in mixed ion crystals of $4^{0}Ca^{2+}$ and $4^{0}Ca^{+}$. Appl Phys B 114:11–16
73. Feldker T et al (2015) Rydberg excitation of a single trapped ion. Phys Rev Lett 115:173001
74. Bachor P, Feldker T, Walz J, Schmidt-Kaler F (2016) Addressing single trapped ions for Rydberg quantum logic. J Phys B 49:154004
75. Feldker T (2017) Rydberg excitation of trapped ions. PhD thesis, Johannes Gutenberg-Universität, Mainz
76. Schmidt-Kaler F et al (2011) Rydberg excitation of trapped cold ions: a detailed case study. New J Phys 13:075014 (2011)

Chapter 2
Properties of Strontium Rydberg Ions

In this chapter the electronic structure of Rydberg ions is described before key properties are reviewed. The relevant length scales in a Rydberg ion system are described in Sect. 2.3, Rydberg state lifetimes are presented in Sect. 2.4, and the interaction strength of microwave (MW)-dressed Rydberg ions and the fidelity of a two-qubit Rydberg gate are discussed in Sect. 2.6.

2.1 Electronic Structure

$^{88}Sr^+$ is a multi-electron system; 37 electrons orbit the nucleus of charge +38. Rydberg states are well-described by a single valence electron orbiting a spherical core with charge $Z = +2$, since the other 36 electrons form closed shells and shield the nuclear charge. Rydberg states of $^{88}Sr^+$ thus follow the Rydberg scaling relations in Table 1.1. Because the valence electron wavefunction extends inside the core, the core electrons do not completely shield the nucleus. As a result the scaling relations are followed in terms of the effective principal quantum number $n^* = n - \mu$, where the quantum defect μ is determined empirically.[1] μ depends on the atomic species as well as the angular momentum quantum numbers L and J. For example the binding energy formula is adapted as follows:

$$E = -\frac{Z^2 R}{n^2} \rightarrow -\frac{Z^2 R}{(n-\mu)^2}, \tag{2.1}$$

where R is the Rydberg constant. Sr^+ energy series are displayed in Fig. 2.1, literature values are used for μ [1]. The $^{88}Sr^+$ $nS_{1/2}$ energy series is measured in this work, the results are in Sect. 4.6.

[1]The symbol δ is commonly used for the quantum defect. In this work μ is used to avoid confusion with the partial derivative symbol in Sect. 4.6.

G. Higgins, *A Single Trapped Rydberg Ion*, Springer Theses,
https://doi.org/10.1007/978-3-030-33770-4_2

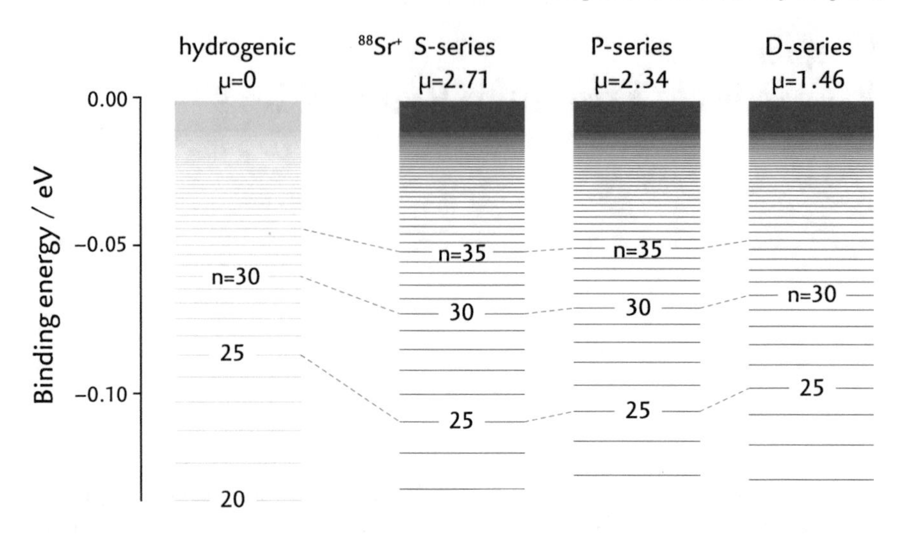

Fig. 2.1 Rydberg state energies depend on the principal quantum number and the quantum defect μ. The energy levels of a hydrogen-like system with $Z = +2$ and of different $^{88}\text{Sr}^+$ series do not match because the quantum defects (μ) are different. Principal quantum numbers (n) are labelled and μ values are taken from [1]

2.2 Atomic Wavefunctions

Atomic wavefunctions describing the multi-electron $^{88}\text{Sr}^+$ ion are derived by reducing the many-body problem to a two-body problem; the inner electrons form closed shells which screen the nuclear Coulomb potential and thus the valence electron experiences a screened Coulomb potential. The screened Coulomb potential is described by parameters from [2]. The Schrödinger equation for the valence electron moving around the core is solved numerically to give atomic wavefunctions. This method is described more completely in [3].

Wavefunctions of $^{88}\text{Sr}^+$ were calculated by Weibin Li at the University of Nottingham and used to produce theoretical values of Rydberg state polarisabilities (Sect. 6.1), quadrupole moments (Sect. 6.2), and natural lifetimes (Sect. 7.4.2), as well as branching ratios of decays from excited states. I calculated similar wavefunctions and Rydberg state properties by adapting the Python package ARC [4] which was developed for calculations of neutral Rydberg atom properties.

2.3 Length Scales

The Rydberg orbital radius scales with n^{*2}, and so Rydberg ions are much bigger than ground state ions. The Rydberg ion experiments (Sect. 1.3.2) and most of the theory proposals (Sect. 1.3.1) involve a string of ions confined in a linear Paul trap

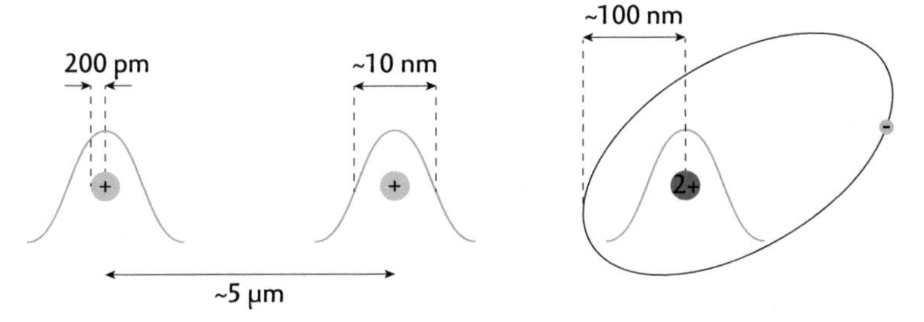

Fig. 2.2 Ground state ions (yellow) are much smaller than the extent of a trapped ion's motion (green). The orbital radius of Rydberg states considered in this thesis is larger than the extent of the ion motion and smaller than the separation between two trapped ions. The Rydberg electron orbital is classically represented. The orbital radius is $\approx$100 nm for $53S_{1/2}$

(Sect. 3.2), with ions separated by $\sim$5 μm. For the Rydberg states considered in this work ($n < 60$), the orbital radius is much less than the inter-ion spacing and thus the overlap of two Rydberg-ion wavefunctions is negligible, as shown in Fig. 2.2.

Each ion is trapped in a 3D pseudo-harmonic potential. Unlike point-like ground-state ions, the Rydberg orbital radius ($\approx$100 nm for $53S_{1/2}$) exceeds the extent of the ion motion ($\sim$10 nm) and trap effects specific to Rydberg ions emerge (Chap. 6).

2.4 State Lifetime

The natural lifetime of a trapped Rydberg ion may be shortened by transitions driven by blackbody radiation and transitions driven by the trapping electric fields.

The natural lifetime of an atomic state m is given by the inverse of the sum of radiative decay rates

$$\tau_{m,\mathrm{nat}} = \left(\sum_{m'} A_{mm'} \right)^{-1}, \tag{2.2}$$

where the Einstein A-coefficient $A_{mm'}$ is the radiative decay rate from m to m'. Radiative decay of Rydberg states is strongest to low-lying states. Higher Rydberg states overlap less with low-lying states and as a result higher Rydberg states have higher natural lifetimes. The natural lifetime scales as n^{*3}, as shown in Fig. 2.3.

Blackbody radiation drives transitions between states and shortens the lifetime of atomic state m [5]

$$\tau_{m,\mathrm{BBR}} = \left(\sum_{m'} A_{mm'} + \sum_{m'} A_{mm'}\bar{n} + \sum_{m''} A_{m''m}\bar{n} \right)^{-1}, \tag{2.3}$$

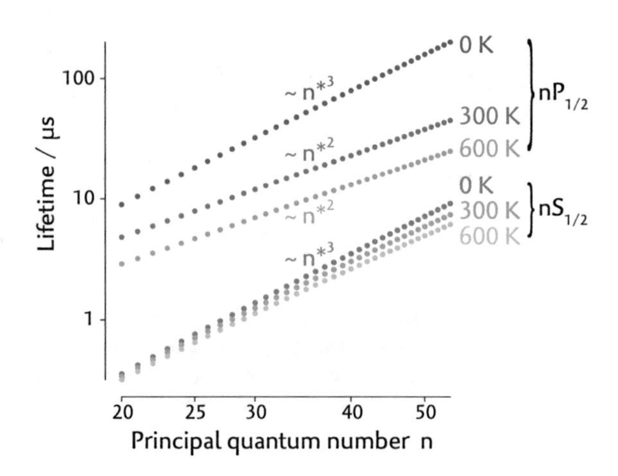

Fig. 2.3 Lifetimes of Rydberg $S_{1/2}$- and $P_{1/2}$-states of $^{88}\mathrm{Sr}^+$ with different temperatures of surroundings. Natural lifetimes scale with n^{*3}, lifetimes limited by blackbody radiation scale with n^{*2}. Trap effects do not affect the lifetimes of the states which are shown. Logarithmic scales are used

where the photon occupation number is

$$\bar{n} = \left(e^{\Delta/kT} - 1\right)^{-1} \tag{2.4}$$

and Δ is the energy difference between the states m and m' or m and m''. While population only moves to states lower in energy (m') by radiative decay, blackbody radiation also drives transitions to higher-energy states (m''). Because of the $\bar{n}$-dependence, blackbody radiation drives transitions between energetically-similar states most strongly. Rydberg state lifetimes scale as n^{*2} if they are limited by blackbody radiation, as shown in Fig. 2.3.

The trapping electric fields strongly drive transitions between Rydberg states with $J > \frac{1}{2}$, since these states have large electric quadrupole moments (see Sect. 6.2). The lifetimes of Rydberg states with $J > \frac{1}{2}$ are thus significantly shortened by trapping electric fields [3]. Transitions driven by typical trapping fields are negligible for $J = \frac{1}{2}$ Rydberg states up to at least $n = 50$ [6]; as a result $J = \frac{1}{2}$ Rydberg states have longer lifetimes, which are limited by natural decay or blackbody radiation driven transitions. Because Rydberg gate fidelities are higher for longer lived Rydberg states [7] the Rydberg ion gate proposals (Sect. 1.3.1) employ $nS_{1/2}$ and $nP_{1/2}$ states (which have $J = \frac{1}{2}$).

As shown in Fig. 2.3 Rydberg states $50S_{1/2}$ and $50P_{1/2}$ have lifetimes $\sim$10 μs when the temperature of the surroundings is 300 K (similar to the laboratory temperature). Coherent Rydberg excitation and Rydberg gates thus require Rabi frequencies $>\sim 2\pi \times 1$ MHz.

In Sect. 7.4.2 the measurement of the $42S_{1/2}$ lifetime is shown.

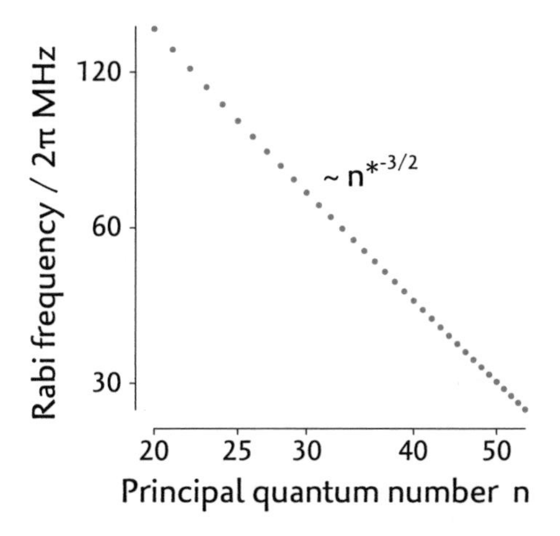

Fig. 2.4 It becomes progressively more difficult to drive a transition between a low-lying state and a Rydberg state as the principal quantum number n is increased. The Rabi frequencies for $6P_{3/2},\ m_J = -\frac{3}{2} \leftrightarrow nS_{1/2},\ m_J = -\frac{1}{2}$ transitions driven by 10 mW of laser light focussed to a 10 μm beam waist are shown. These transitions are constituents of the two-photon transitions used in the experiment. The plotted Rabi frequencies use calculated dipole moments of the $6P_{3/2},\ m_J = -\frac{3}{2} \leftrightarrow nS_{1/2},\ m_J = -\frac{1}{2}$ transition. Logarithmic scales are used

2.5 Rydberg Excitation Rabi Frequency

As the principal quantum number is increased the overlap of a Rydberg state with a low-lying state is reduced, and as a result the transition dipole moment between a low-lying state and a Rydberg state becomes smaller. The transition dipole moment and the Rydberg-excitation Rabi frequency scale as $n^{*-3/2}$, as shown in Fig. 2.4.

Provided the Rydberg state decay rate is limited by natural decay or by blackbody radiation-induced transitions, coherent excitation of higher Rydberg states may be achieved with lower laser light intensities, due to difference in scaling between the Rydberg excitation Rabi frequency ($n^{*-3/2}$) and the Rydberg state decay rate (n^{*-3} for natural decay or n^{*-2} for blackbody radiation-induced decay).

2.6 Two-Qubit Rydberg Gate

The Rydberg ion gate proposals (Sect. 1.3.1) involve coupling $nS_{1/2}$ and $nP_{1/2}$ states using MW radiation. The coupled states have dipole moments, with magnitudes which scale as n^{*2}. Orbital sizes and the transition dipole moment $\langle nLJ|er|nL'J'\rangle$ scale in the same fashion. The strength of the dipole-dipole interaction between two MW-dressed Rydberg states thus scales as n^{*4}, as shown in Fig. 2.5a. The interaction strength also depends on the separation between the ions Δz; the strength varies as Δz^{-3}.

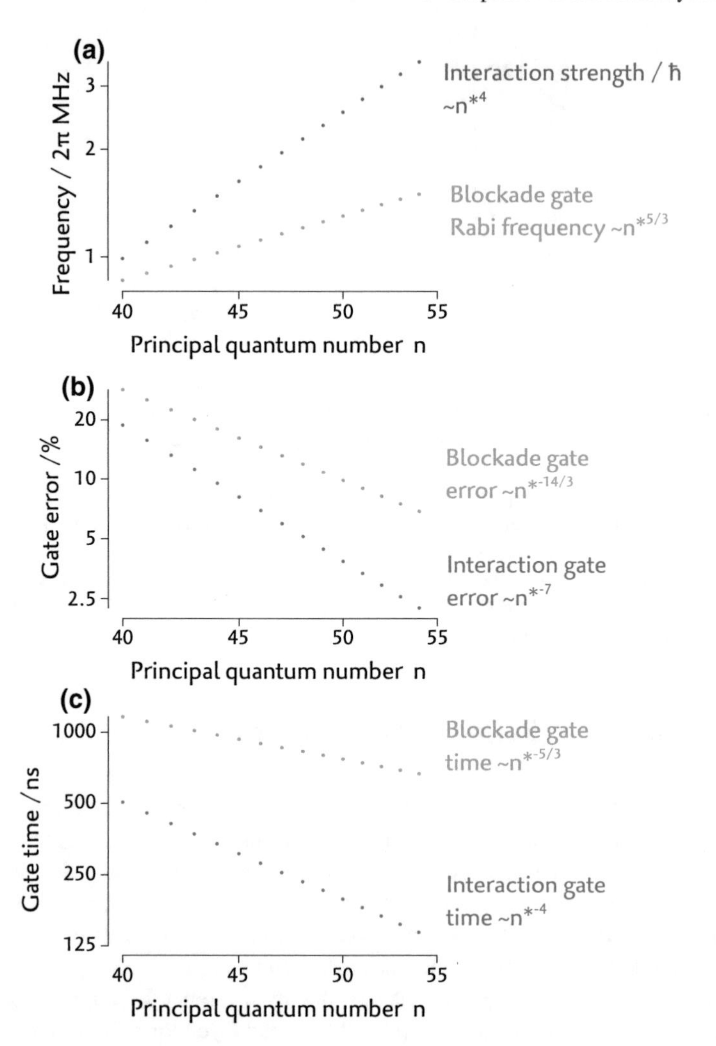

Fig. 2.5 Higher Rydberg states interact more strongly and can be used for faster gates with lower errors. **a** The dipole-dipole interaction strength is calculated for two MW-dressed ions separated by 4 μm; the rest of the calculations use the same ion separation. For the blockade gate, higher Rydberg states require higher Rydberg excitation Rabi frequencies. **b** The gate errors are lower for higher Rydberg states with stronger interactions and longer lifetimes; the states considered here have lifetimes limited by natural decay. **c** Sub-microsecond gate times may be obtained. Logarithmic scales are used

Two-qubit Rydberg gates using the dipole-dipole interaction are more efficient when higher Rydberg states with stronger interactions and longer lifetimes are used,[2] as shown in Fig. 2.5b. A Rydberg blockade gate is described in [7]; the gate error is proportional to $(U\tau_r)^{-2/3}$, where U is the interaction strength and τ_r is the Rydberg state lifetime. This gate requires higher Rydberg excitation Rabi frequencies for higher Rydberg states; the optimal Rabi frequency is proportional to $\sqrt[3]{\frac{U^2}{\tau}}$. This means higher laser light intensities are required when higher Rydberg states are used for the blockade gate. A Rydberg interaction gate is described in [8]; the gate error is proportional to $(U\tau_r)^{-1}$. The future implementation of such a gate in our system is discussed at the end of Sect. 7.4.3.

These gates may be carried out over sub-microsecond timescales, as shown in Fig. 2.5c. The time required for the blockade gate is inversely proportional to the Rydberg excitation Rabi frequency, while the time required for the interaction gate is inversely proportional to the interaction strength.

References

1. Lange V, Khan MA, Eichmann U, Sandner W (1991) Rydberg states of the strontium ion. Z Phys D 18:319–324
2. Aymar M, Greene CH, Luc-Koenig E (1996) Multichannel Rydberg spectroscopy of complex atoms. Rev Mod Phys 68:1015–1123
3. Schmidt-Kaler F et al (2011) Rydberg excitation of trapped cold ions: a detailed case study. New J Phys 13:075014
4. Šibalić N, Pritchard J, Adams C, Weatherill K (2017) ARC: an open-source library for calculating properties of alkali Rydberg atoms. Comput Phys Commun 220:319–331
5. Gallagher TF, Cooke WE (1979) Interactions of blackbody radiation with atoms. Phys Rev Lett 42:835–839
6. Müller M, Liang L, Lesanovsky I, Zoller P (2008) Trapped Rydberg ions: from spin chains to fast quantum gates. New J Phys 10:093009
7. Saffman M, Walker TG, Mølmer K (2010) Quantum information with Rydberg atoms. Rev Mod Phys 82:2313–2363
8. Jaksch D et al (2000) Fast quantum gates for neutral atoms. Phys Rev Lett 85:2208–2211

[2]Higher principal quantum number states have longer lifetimes provided they are limited by natural decay or blackbody radiation-induced transitions.

Chapter 3
Experimental Setup

This chapter deals with the main parts of the laboratory and how they come together for experiments to be carried out. During experiments trapped ions are manipulated using laser pulses. In Sect. 3.1 general experimental sequences are described and experimental requirements are established. In the subsequent sections systems developed to meet these requirements are presented, namely the linear Paul trap (Sect. 3.2), laser systems (Sect. 3.3) and electronic control systems (Sect. 3.4).

3.1 Experiments with a Single Trapped Ion

A single trapped ion is probed in most of the experiments in this thesis. Experiments involve sequences of laser pulses which drive transitions between states of an ion. The laser-driven transitions are described in Sect. 3.1.1, thereafter the key steps in an experiment are introduced. A typical experiment consists of the following steps:

1. Laser cooling
2. State initialisation
3. Transitions are driven by the Rydberg-excitation lasers or the qubit laser
4. Measurement.

The laser cooling, state initialisation and measurement steps are described in Sects. 3.1.2, 3.1.3 and 3.1.4 respectively.

3.1.1 Laser-Driven Transitions

Within a linear Paul trap ions are confined in effective harmonic trapping potentials in the three spatial dimensions. The trapping potentials are described in more detail

G. Higgins, *A Single Trapped Rydberg Ion*, Springer Theses,
https://doi.org/10.1007/978-3-030-33770-4_3

in Sect. 6.1.1. Trapped ions have electronic degrees of freedom and motional degrees of freedom; laser light is used to drive transitions between electronic and motional states.

Electronic States of $^{88}Sr^+$

The eigenenergies of different electronic states are represented in the level scheme in Fig. 3.1. A magnetic field splits the levels into Zeeman sublevels, which are labelled by the magnetic quantum number m_J (see Sect. 3.2.6).

A Zeeman sublevel of the ground state $5S_{1/2}$ and a Zeeman sublevel of $4D_{5/2}$ are used to store an ion qubit. Electric quadrupole transitions between $5S_{1/2}$ and $4D_{5/2}$ sublevels are driven by 674 nm laser light with Rabi frequencies $\sim 2\pi \times 100$ kHz. These Rabi frequencies greatly exceed the $4D_{5/2}$ natural decay rate $2\pi \times 410$ mHz [3], and thus qubit coherence times can greatly exceed quantum gate times. $5S_{1/2} \leftrightarrow 4D_{5/2}$ transitions are also used for resolved sideband cooling (Sect. 3.1.2), optical pumping (Sect. 3.1.3), micromotion compensation (Sect. 3.2.8) and ion temperature measurements (described later in this section).

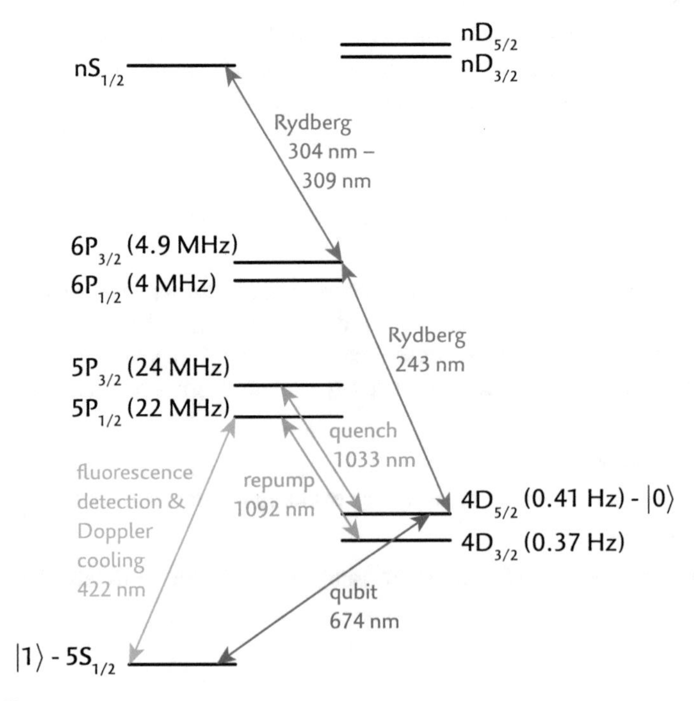

Fig. 3.1 $^{88}Sr^+$ level scheme including natural decay rates and the lasers used for driving transitions. Natural decay rates $\Gamma/2\pi$ are taken from [1–4] and the $6P_{3/2}$ decay rate is measured in Sect. 4.1.1. Zeeman sublevels are not shown

The rest of the lasers are used to drive electric dipole transitions. The $2\pi \times 22$ MHz natural decay rate of $5P_{1/2}$ [4] means photons are scattered at a high rate when the $5S_{1/2} \leftrightarrow 5P_{1/2}$ transition is strongly driven by 422 nm laser light. Owing in part to this, the $5S_{1/2} \leftrightarrow 5P_{1/2}$ transition is used for ion measurement (which relies on detection of scattered photons, see Sect. 3.1.4) as well as Doppler cooling (Sect. 3.1.2).

When the 422 nm laser is employed 1092 nm laser light is often used to prevent optical pumping to $4D_{3/2}$, which is caused by decay from $5P_{1/2}$ to $4D_{3/2}$ in 6% of cases [5]. 1033 nm laser light is used to remove population from $4D_{5/2}$ and initialise the system (Sect. 3.1.3), it is also used for carrying out resolved sideband cooling.

Laser light at 243 and 304–309 nm is used to drive two-photon transitions from either of the metastable $4D$ states to Rydberg S- or D-states (Chap. 4).

Laser-Driven Transitions Between Electronic and Motional States

Spectra include carrier lines which are flanked by sidebands. A carrier line corresponds to a transition in which the electronic state is changed and the number of phonons is kept the same. When a sideband transition is excited the electronic state and the number of phonons are changed. Blue sidebands are more energetic and correspond to transitions in which the phonon number is increased, red sidebands are less energetic and correspond to transitions in which the phonon number is decreased. Phonon-number-changing transitions occur because of the momentum kick imparted to the trapped ion when a photon is absorbed. These transitions are shown in Fig. 3.2.

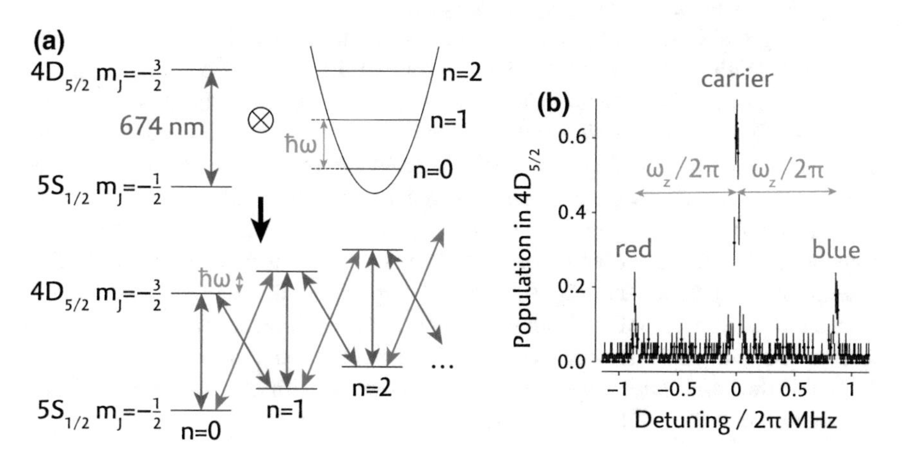

Fig. 3.2 **a** Laser light drives transitions between electronic states and motional states. **b** Spectrum showing the $5S_{1/2},\ m_J = -\frac{1}{2} \leftrightarrow 4D_{5/2},\ m_J = -\frac{3}{2}$ carrier transition resonance and axial sidebands. The axial frequency is $\omega_z = 2\pi \times 867$ kHz. Only first-order sidebands are shown in (**a**) and (**b**). Error bars indicate quantum projection noise (68% confidence interval)

The relative strength with which phonon-number-changing transitions are driven depends on the ratio of the recoil energy to the spacing of the quantum harmonic oscillator energy levels. This is described by the Lamb-Dicke parameter η [6]

$$\begin{aligned} \eta^2 &= \left(\frac{\hbar^2 k^2 \cos^2\phi}{2M}\right) \Big/ \hbar\omega \\ &= \frac{\hbar k^2 \cos^2\phi}{2M\omega}. \end{aligned} \tag{3.1}$$

Here M is the ion mass, ω is the frequency of a particular motional mode, k is the laser light wavevector and ϕ is the angle between the laser beam and the mode axis. When the recoil energy is much smaller than the quantised energy spacing, the Lamb-Dicke parameter is small and phonon-number-changing transitions are driven less strongly. The Lamb-Dicke parameter is different for different transitions, as well as different motional modes.

The relative strength with which sideband transitions and carrier transitions are driven depends not only on the Lamb-Dicke parameter, but also on the number of phonons in a particular mode, n.[1] The Lamb-Dicke regime is defined by the inequality

$$\eta^2(2n+1) \ll 1. \tag{3.2}$$

Within the Lamb-Dicke regime transitions which change the motional quantum number by more than one are strongly suppressed.

Working within the Lamb-Dicke regime is required for transitions to be driven coherently with high efficiency (since unwanted driving of sideband transitions causes dephasing), for resolved sideband cooling (which requires the phonon number to be usually preserved during spontaneous decay, see Sect. 3.1.2) and it is advantageous for high-precision spectroscopy (to avoid Doppler broadening of resonance lines). The Lamb-Dicke parameter for the axial motional mode and a 674 nm laser beam (used to drive the qubit transition) at 45° to the trap axis is 0.05 when typical trapping parameters are used (for typical trapping parameters see Sect. 3.2.1). This laser beam is represented in Fig. 3.3a. After Doppler cooling (Sect. 3.1.2) an ion has $\approx$16 phonons in the axial mode and the Lamb-Dicke regime inequality [Eq. (3.2)] is satisfied for this particular laser beam and the axial motional mode.

Within the Lamb-Dicke regime the relative strengths with which carrier transitions and first-order sideband transitions are driven are given by [6]

$$\begin{aligned} \Omega_{n\to n} &= \left(1-\eta^2 n\right)\Omega_0, \\ \Omega_{n\to n+1} &= \eta\sqrt{n+1}\,\Omega_0, \\ \Omega_{n\to n-1} &= \eta\sqrt{n}\,\Omega_0, \end{aligned} \tag{3.3}$$

[1] This n-dependence can be understood in terms of the n-dependence of the ladder operators $\hat{a}$ and $\hat{a}^\dagger$.

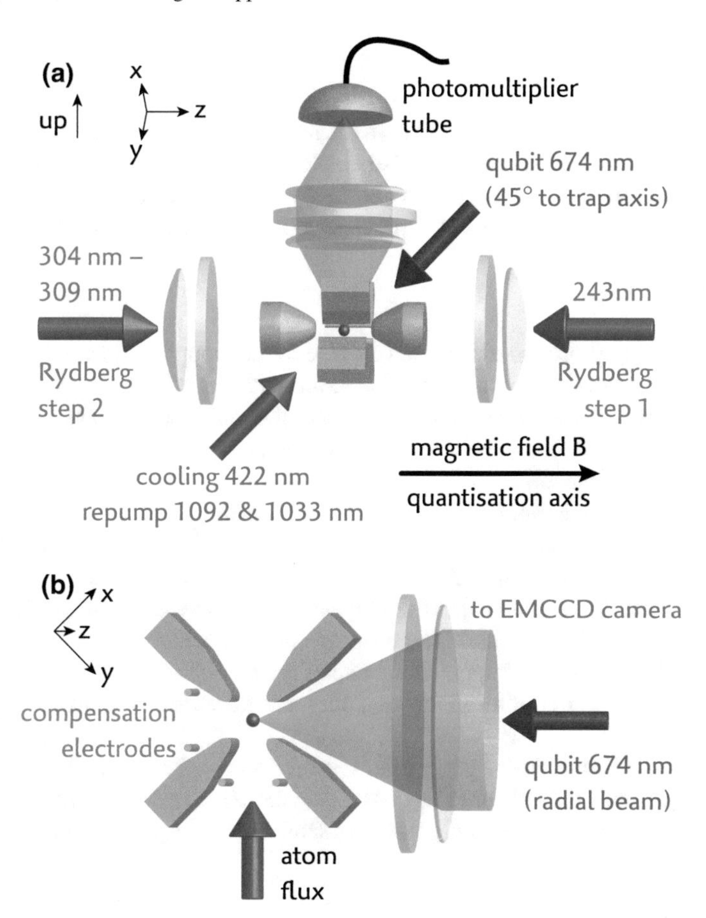

Fig. 3.3 An ion is trapped between the linear Paul trap 'blade' and 'endcap' electrodes, which are gold-coated. Laser beams enter the vacuum chamber through viewports; the UV lasers pass through holes in the endcaps. Each micromotion compensation electrode consists of a pair of steel rods (shown in silver). Ion fluorescence is collected on the PMT and the EMCCD camera. Ions are loaded by resistivity-heating a Sr source: a flux of Sr atoms passes through the trap and lasers collinear with the 243 nm laser photo-ionise atoms. **a** Side view, **b** along the trap axis

written to second order in $(\eta\sqrt{n})$; Ω_0 is the Rabi frequency of carrier transitions when $n = 0$. The ion temperature, that is the mean phonon number, is determined by comparing the Rabi frequencies with which sideband transitions are driven [6].

Two-photon transitions between $4D$ states and Rydberg states are described by an effective Lamb-Dicke parameter, which is introduced in Sect. 4.4.

3.1.2 Laser Cooling

At the beginning of each experiment run the ion is Doppler cooled. In many experiments resolved sideband cooling is also employed. Doppler cooling is used to cool an ion to the Lamb-Dicke regime[2] while resolved sideband cooling is used to cool an ion below the Doppler cooling limit close to the motional ground state.

Doppler Cooling

Doppler cooling proceeds as follows [6]: Due to the Doppler effect, the absorption coefficient of a laser-driven transition depends on the ion velocity. If a laser is appropriately detuned from the transition resonance, the radiation pressure force exerted by the laser light on the ion behaves as a drag force. The kinetic energy of the ion is then reduced and the ion is cooled.

In our lab the ion is Doppler cooled using 422 nm laser light red-detuned from the $5S_{1/2} \leftrightarrow 5P_{1/2}$ resonance. The 422 nm laser beam k-vector has components along all three of the trap axes [Fig. 3.3a], this allows Doppler cooling in all directions with a single 422 nm laser beam [6]. After Doppler cooling the mean phonon number in the axial mode is $\bar{n}_z \approx 16$ and the mean phonon numbers in the radial modes are $\bar{n}_x, \bar{n}_y \approx 12$ when typical trapping frequencies are used. The mean phonon numbers dictate the temperatures of the motional modes, and they are determined as described in Sect. 3.1.1.

1033 and 1092 nm laser light is used together with 422 nm laser light; 1092 nm laser light prevents optical pumping to $4D_{3/2}$, while 1033 nm laser light removes any ion population which was initially in $4D_{5/2}$. Doppler cooling is typically carried out for 1 ms.

Resolved Sideband Cooling

Resolved sideband cooling is a process in which ion population is optically pumped close to the motional ground state [6]. The process proceeds as follows (schematic in Fig. 3.4): The red sideband of a $5S_{1/2} \leftrightarrow 4D_{5/2}$ transition is driven; each excitation to $4D_{5/2}$ reduces the phonon number by 1. From $4D_{5/2}$ population may be driven back to $5S_{1/2}$ (and the lost phonon is regained) or population may decay to $5S_{1/2}$. Within the Lamb-Dicke regime the decay usually preserves the phonon number and thus on average the phonon number is reduced during the process. The process is sped up by coupling $4D_{5/2}$ and $5P_{3/2}$ such that the effective lifetime of $4D_{5/2}$ is reduced.

The radial 674 nm laser beam in Fig. 3.3 is used to cool radial motional modes, the angled beam is used to cool the axial motional mode. Typically sideband cooling is carried out for 1 ms for the axial mode and for 1 ms for the two radial modes. After sideband cooling the mean phonon numbers in the radial modes and in the axial mode are $\bar{n}_x,\ \bar{n}_y,\ \bar{n}_z \approx 0.2$ when typical trapping frequencies are used.

---

[2]With regards the transitions driven in the experiment and the typical trapping frequencies used.

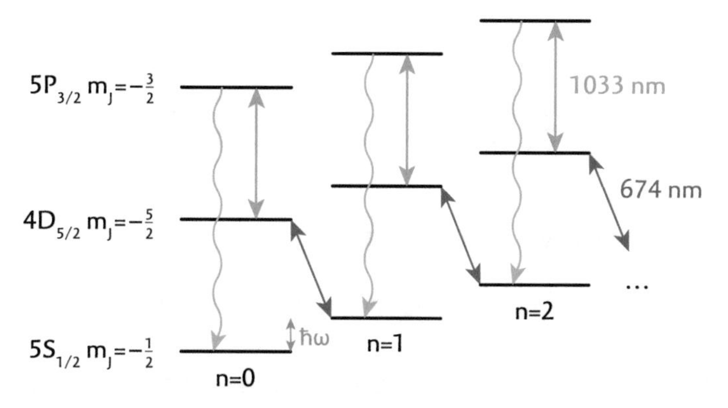

Fig. 3.4 During sideband cooling each excitation from $5S_{1/2}$ to $4D_{5/2}$ reduces the phonon number (n) by 1, and decay to $5S_{1/2}$ usually preserves the phonon number (within the Lamb-Dicke regime). The effective lifetime of the excited state is shortened by coupling $4D_{5/2}$ and $5P_{3/2}$. The energy spacing of the motional levels is $\hbar\omega$

3.1.3 State Initialisation

Different experiments require the ion to be initialised in different states. The initialisation procedures are described in this section.

Initialisation in $|1\rangle \equiv 5S_{1/2},\ m_J = -\frac{1}{2}$ by Frequency-Resolved Optical Pumping

There are two $5S_{1/2}$ states, labelled by $m_J = \pm\frac{1}{2}$. State $5S_{1/2},\ m_J = -\frac{1}{2} \equiv |1\rangle$ is used together with $4D_{5/2},\ m_J = -\frac{5}{2} \equiv |0\rangle$ to store a qubit. Initialisation in $|1\rangle$ proceeds as follows: First, any ion population residing in the metastable $4D$ states is removed and sent to $5S_{1/2}$ via the $5P$ states using 1033 and 1092 nm laser light. Then any ion population residing in $5S_{1/2},\ m_J = +\frac{1}{2}$ is optically pumped to qubit state $|1\rangle \equiv 5S_{1/2},\ m_J = -\frac{1}{2}$ by driving the following cycle ~10 times:

1. A π pulse is applied on the frequency-resolved $5S_{1/2},\ m_J = +\frac{1}{2} \leftrightarrow 4D_{5/2},\ m_J = -\frac{3}{2}$ transition. Any population residing in $|1\rangle$ is unaffected, while population in $5S_{1/2},\ m_J = +\frac{1}{2}$ is excited to $4D_{5/2},\ m_J = -\frac{3}{2}$.
2. Population is removed from the $4D$ states and returned to $5S_{1/2}$ via the $5P$ states using 1033 and 1092 nm laser light.

Initialisation in $|0\rangle \equiv 4D_{5/2},\ m_J = -\frac{5}{2}$

The ion is prepared in $|1\rangle$ by optical pumping and then excited to qubit state $|0\rangle \equiv 4D_{5/2},\ m_J = -\frac{5}{2}$ by applying a π pulse on the 674 nm $|1\rangle \leftrightarrow |0\rangle$ transition. Electron shelving (Sect. 3.1.4) is then used to confirm excitation to $|0\rangle$; cases in which initialisation is unsuccessful are removed during data analysis.

Initialisation in $|1\rangle + e^{i\phi}|0\rangle$

The ion is prepared in $|1\rangle$ by optical pumping and then the superposition $|1\rangle + e^{i\phi}|0\rangle$ is excited by driving a $\frac{\pi}{2}$ pulse on the 674 nm $|1\rangle \leftrightarrow |0\rangle$ transition with laser phase $\phi + \frac{\pi}{2}$.

Initialisation in a Phonon Number State

The ion is prepared in a state with n_x and n_y radial phonons in the x and y radial modes as follows:

1. The radial motional modes are sideband cooled close to their ground states.
2. The ion is prepared in $|1\rangle$ by optical pumping.
3. The following cycle is carried out n_x times:
   (a) A π pulse is applied on the x-motional mode blue sideband of the $|1\rangle \leftrightarrow |0\rangle$ transition. The pulse length accounts for the phonon-number-dependent Rabi frequency described by Eq. (3.3).
   (b) Electron shelving is used to confirm excitation to $|0\rangle$.
   (c) A π pulse is applied on the $|1\rangle \leftrightarrow |0\rangle$ carrier transition to return population to $|1\rangle$.
   (d) 1033 and 1092 nm laser light is used to remove any residual population from the $4D$ states.
4. The cycle is repeated n_y times, this time the y-motional mode blue sideband is driven.
5. In some experiments the ion is then transferred to $|0\rangle$ by driving a π pulse on the $|1\rangle \leftrightarrow |0\rangle$ transition.

This method allows us to prepare an ion with 40 radial phonons with $\approx$20% efficiency. We check the initialisation by preparing an ion in $|1\rangle$ with n_y phonons in the y-mode and then driving Rabi oscillations on the y-motional mode blue sideband of the $|1\rangle \leftrightarrow |0\rangle$ transition, results are shown in Fig. 3.5. The Rabi frequency scales as $\sqrt{n_y + 1}$, as expected from Eq. (3.3).

Initialisation in $4D_{3/2}$

Population is optically pumped from $5S_{1/2}$ to $4D_{3/2}$ by driving the 422 nm $5S_{1/2} \leftrightarrow 5P_{1/2}$ transition; from $5P_{1/2}$ the ion decays to $4D_{3/2}$ in 6% of the cases [5]. A mixture of $4D_{3/2}$ Zeeman sublevels is populated.

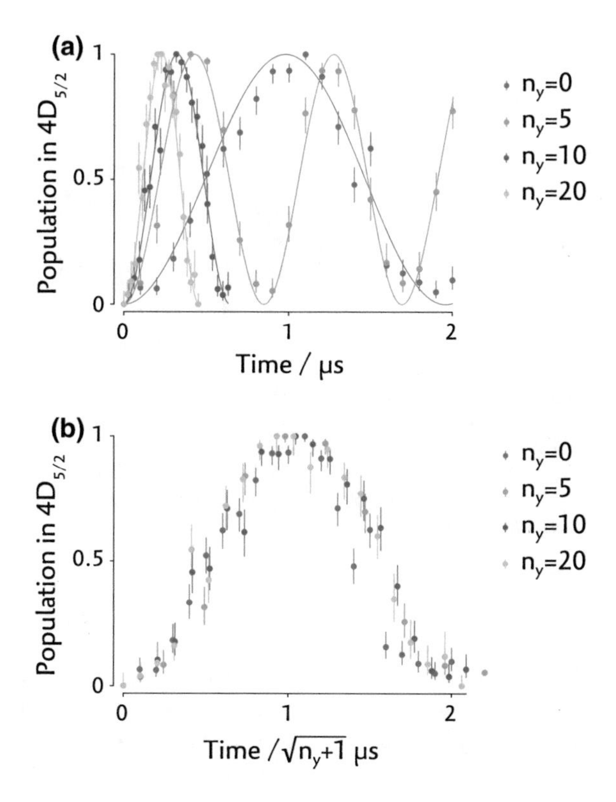

Fig. 3.5 Preparation of phonon number states is demonstrated by driving high-contrast Rabi oscillations on a $5S_{1/2} \leftrightarrow 4D_{5/2}$ blue motional sideband transition and observing the $\sqrt{n_y+1}$-scaling of the Rabi frequency with the phonon number n_y in the y radial mode. **a** The horizontal axis is the excitation time, **b** the time axis is scaled to make the $\sqrt{n_y+1}$-dependence clear. Sinusoidal fit functions in (**a**) serve to guide the eye. Error bars indicate quantum projection noise (68% confidence interval)

3.1.4 Measurement

Measurements involve a technique called electron shelving. Electron shelving distinguishes an ion in state $4D_{5/2}$ from an ion in states $5S_{1/2}$ or $4D_{3/2}$, as described below.

In terms of a qubit stored in $5S_{1/2}$ and $4D_{5/2}$, electron shelving is a projective measurement in the $\hat{\sigma}_z$ basis. By driving a $\frac{\pi}{2}$ pulse on the $|1\rangle \leftrightarrow |0\rangle$ transition with laser phase 0 ($\frac{\pi}{2}$) before carrying out electron shelving, a projective measurement in the $-\hat{\sigma}_y$ ($-\hat{\sigma}_x$) basis is carried out. Measurements in different bases allow us to carry out quantum state tomography and quantum process tomography, these techniques are described in Sect. 7.4.3.

To distinguish population in $5S_{1/2}$ and $4D_{3/2}$ we first transfer population from the $5S_{1/2}$ sublevels to $4D_{5/2}$ sublevels by driving 674 nm transitions and then we use electron shelving.

Electron Shelving

The ion is illuminated by 422 and 1092 nm laser light and we look for scattered 422 nm photons using a PMT. When the PMT detects a photon it sends a digital

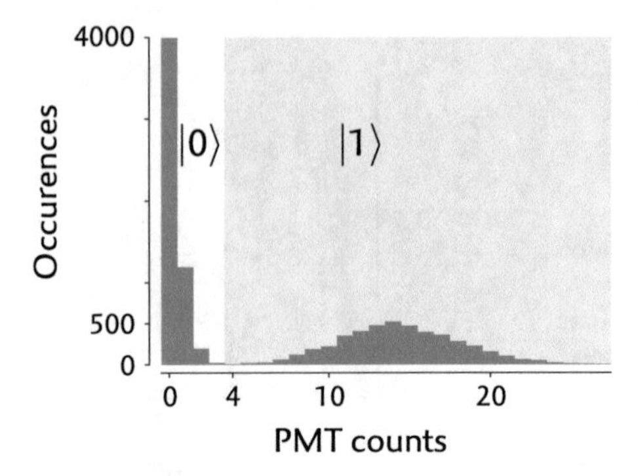

Fig. 3.6 Electron shelving using photon counts measured with the PMT. If 4 or more counts are recorded during the 500 μs measurement time the ion is most probably in state $5S_{1/2}$ (qubit state $|1\rangle$). If 3 or fewer counts are recorded the ion is likely in state $4D_{5/2}$ (qubit state $|0\rangle$). The data is described well by two Poissonian distributions with mean values 0.3 and 14, the overlap of the distributions is 0.04%. This overlap as well as decay from $4D_{5/2}$ during the measurement time cause errors $\sim$0.1%

signal to a counter, which is connected to a PC (Sect. 3.4). If the ion was initially in $5S_{1/2}$ or $4D_{3/2}$ 422 nm photons are scattered by the ion, whereas if the ion is in state $4D_{5/2}$ no light is scattered by the ion and the PMT shows only a background signal due to background light and PMT dark counts. Measuring for 500 μs allows us to distinguish between an ion in state $4D_{5/2}$ and an ion in state $5S_{1/2}$ or $4D_{3/2}$ with $\sim$99.9% efficiency, as shown in Fig. 3.6. The 500 μs detection time greatly exceeds the lifetimes of all the atomic states except for the metastable $4D$ states (lifetimes $\sim$400 ms [1, 3]) and the ground state $5S_{1/2}$.

The states of multiple ions are measured by imaging the ions on a camera and using the electron shelving technique. However, most of the measurements in this thesis are concerned with a single ion.

3.2 Linear Paul Trap

In this experiment ions are trapped in a macroscopic linear Paul trap in an ultra-high vacuum ($<10^{-10}$ mbar). In this section key features of the trap are presented. Information about the assembly of the trap and the vacuum chamber may be found in the master's thesis of Fabian Pokorny [7].

3.2.1 Ions in the Trapping Potential

The linear Paul trap uses a combination of static and oscillating electric quadrupole fields which result in effective harmonic trapping potentials in the three spatial

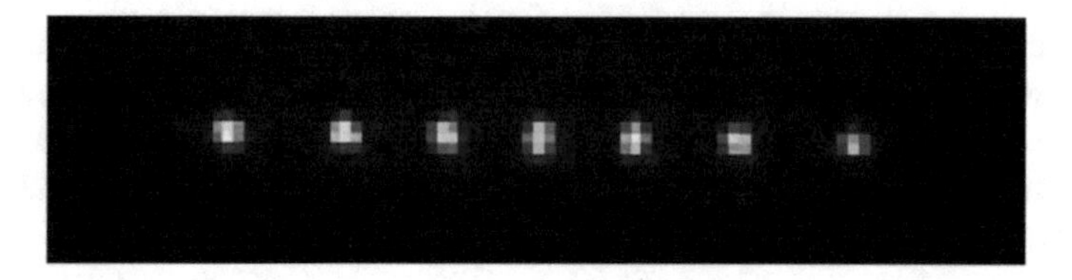

Fig. 3.7 String of seven $^{88}Sr^+$ ions imaged on the EMCCD camera

dimensions, as described in Sect. 6.1.1. The oscillating field is driven at $\Omega_{rf} = 2\pi \times 18.2\,\text{MHz}$. Typically the oscillating field has gradient $\alpha \approx 8.4 \times 10^8\,\text{V m}^{-2}$ and the static field has gradient $\beta \approx 6.8 \times 10^6\,\text{V m}^{-2}$. Trapping frequencies $\omega_{x,y} \approx 2\pi \times 1.7\,\text{MHz}$ and $\omega_z \approx 2\pi \times 840\,\text{kHz}$ result, where x and y are the radial directions and z is the axial direction of the trap (see Fig. 3.3). The confinement is weaker along the trap axis such that multiple trapped ions form a string; the mutual repulsion between the ions keeps them apart by $\approx 5\,\mu\text{m}$.

Ion fluorescence is collected by a PMT[3] and it is also used to image individual ions on an EMCCD camera,[4] as shown in Fig. 3.7.

3.2.2 *Trap Components*

The trap parts were fabricated at the Institute for Quantum Optics and Quantum Information (IQOQI), Innsbruck. The linear Paul trap consists of titanium electrodes mounted in a sapphire holder, which itself is in a titanium mount. The electrodes are coated in gold. A trapped ion is separated from the blade electrodes by $500\,\mu\text{m}$. Holes in the endcap electrodes provide optical access along the trap axis with numerical aperture 0.058. This design builds upon previous macroscopic linear Paul traps machined in Innsbruck.

3.2.3 *Trap Temperature*

In a trap of the same design it was found that the typical powers used for driving the oscillating trapping field (≈ 1 W) increase the trap temperature by $\approx 1.4\,°\text{C}$ [8]. Temperature increases two orders of magnitude higher have been recorded elsewhere [9] resulting in increased blackbody radiation which causes shifts in atomic clock transitions. In our experiment increased blackbody radiation would decrease Rydberg state lifetimes (see Sect. 2.4).

[3]Hamamatsu Photonics H10682-210.

[4]Andor iXon3 897.

The temperature change measured in the trap of the same design as ours is relatively small because of the high thermal conductivity of the trap components and the high electrical conductivity of the gold-coated electrodes. We use similar powers for driving the oscillating trapping field in our experiment and we expect similar temperature changes ($\sim$1.4 °C).

3.2.4 Photo Electrons Induced by UV Light

The gold coating on the electrodes also means photo electrons are unlikely to be produced by stray UV laser light illuminating an electrode surface, owing to the high work function of gold >5.1 eV [10], cf. 243 nm photons carry 5.1 eV energy.

Photo electrons cause the build-up of stray charges inside the chamber which disturb the trapping potential (Sect. 3.2.8), additionally a flux of photo electrons may perturb highly-sensitive Rydberg states [11].

In one of the traps used in the Rydberg ion experiment in Mainz stray charges were generated by VUV laser light hitting surfaces; these charges disturbed the trapping potential and this contributed to their broad Rydberg resonance linewidths [12] (see Sect. 1.3.2).

3.2.5 Motional Heating

Motional heating is detrimental to quantum manipulation of trapped ions [13] and is likely harmful in trapped Rydberg ions systems since the energy for Rydberg excitation depends on the number of phonons in the system (Sect. 6.1). Furthermore motional heating is itself a symptom of electric field noise which may perturb Rydberg ions directly.

The upper bounds of heating rates in our trap are relatively low; in the axial mode the rate is <18 phonon s^{-1} and in the radial modes the rate is <4 phonon s^{-1} when $\omega_{x,y} \approx 2\pi \times 1.7$ MHz and $\omega_z = 2\pi \times 870$ kHz. The heating rate is inferred by first cooling the ion and then measuring the ion temperature (Sect. 3.1.1) after different time delays, the same method is used in [14]. The heating rate is not precisely determined because servo bumps of the qubit laser drive off-resonant transitions making temperature measurements difficult for phonon numbers $< \sim 0.2$.

3.2.6 Magnetic Field

Coils outside the chamber apply a magnetic field along the trap axis with strength $\sim$0.3 mT at the position of trapped ions. Since the magnetic quantum number is a good quantum number in our system (for low-lying states and Rydberg S-states), the field defines the quantisation axis. The magnetic field is aligned along the trap axis as follows: A beam of circularly-polarised 422 nm laser light is sent along the trap axis through holes in the endcap electrodes. When the magnetic field direction is collinear with the laser beam propagation the 422 nm laser light drives only one σ-transition between $5S_{1/2}$ and $5P_{1/2}$ sublevels, and a sublevel of $5S_{1/2}$ becomes a dark state. Currents in the coils are varied until a dark state appears and the ion fluorescence is minimised.

3.2.7 *Ion Loading*

In our experiment a trapped $^{88}Sr^{+}$ ion is lost by double ionisation after typically several hundred excitations to a Rydberg state (see Chap. 5). Double ionisation occurs once in every $\sim$300 excitations to a Rydberg state in the Mainz experiment [11]. Rydberg ion experiments thus require fast and reliable ion loading.

Until recently, and for most of the experiments in this thesis, single $^{88}Sr^{+}$ ions were loaded by resistively heating a strontium oven to produce a beam of strontium atoms which passes through the trap, atoms were then ionised in a two-step photoionisation process [15] and the ions were Doppler cooled (see Sect. 3.1.2). Around 10 min were spent loading a $^{88}Sr^{+}$ ion. After loading we let the oven cool for an additional 10 min before carrying out Rydberg experiments, since we suspect blackbody radiation increases the likelihood of losing a Rydberg ion by double-ionisation.

Recently laser ablation loading was introduced to our system and now ions are loaded in less than 30 s without the system being heated.

3.2.8 *Micromotion Compensation*

In a linear Paul trap ions are confined in a combination of oscillating and static electric quadrupole fields. Ideally the nulls of the electric quadrupole fields overlap and the ion is trapped at the null where the electric field strength is zero. Imperfections in the construction of the trap and stray charges (Sect. 3.2.4) cause the nulls of the quadrupole fields to be at different positions. This results in oscillating and static electric dipole fields at the average position of the ion which drive motion called 'excess micromotion' [16, 17].

Excess micromotion alters atomic transition lineshapes, disrupts laser cooling and shifts resonance frequencies [16, 17]. In quantum information processing experiments and atomic clock experiments steps are taken to minimise excess micromotion. Rydberg ions have extremely large polarisabilities; electric fields cause Rydberg resonance lineshapes to be altered (see Sect. 6.1) and they can also cause field ionisation [18]. Minimisation of micromotion and thus the electric fields experienced by the Rydberg ion is important in our experiment.

Our linear Paul trap has electrodes for minimisation of radial micromotion [see Fig. 3.3b]. By applying appropriate static voltages to these electrodes the position of the null of the static electric quadrupole field is made to overlap with the null of the oscillating electric quadrupole field.

We find appropriate voltages for minimisation of micromotion using established techniques [16, 17]; we monitor the displacement of a trapped ion as the trap depth is changed by imaging the ion on the EMCCD camera, we use the cross-correlation technique, and we employ the resolved sideband method using the $5S_{1/2} \leftrightarrow 4D_{5/2}$ transition. After minimising micromotion using the resolved sideband method the typical strength of the residual oscillating field at the position of the ion is $\sim 10\,\mathrm{V\,m^{-1}}$ when typical trapping frequencies are used.

In Sect. 6.1.2 we investigate the effects of non-overlapping electric quadrupole trapping fields on a Rydberg ion and introduce a novel method for minimising micromotion using a trapped Rydberg ion.

3.3 Laser Systems

Laser systems are described in this section. Laser beams pass from the laser sources (Sect. 3.3.1), through acousto-optic modulators AOMs used for laser pulse shaping (Sect. 3.3.2) before they are brought close the to the experiment chamber via single-mode optical fibres (Sect. 3.3.3). Laser pulses then enter the chamber through viewports (Fig. 3.3) and are focussed onto trapped ions.

So that the transitions in Fig. 3.1 may be driven efficiently laser frequencies are stabilised by using optical resonators as frequency references (Sect. 3.3.4). A schematic of a typical beam path is shown in Fig. 3.8.

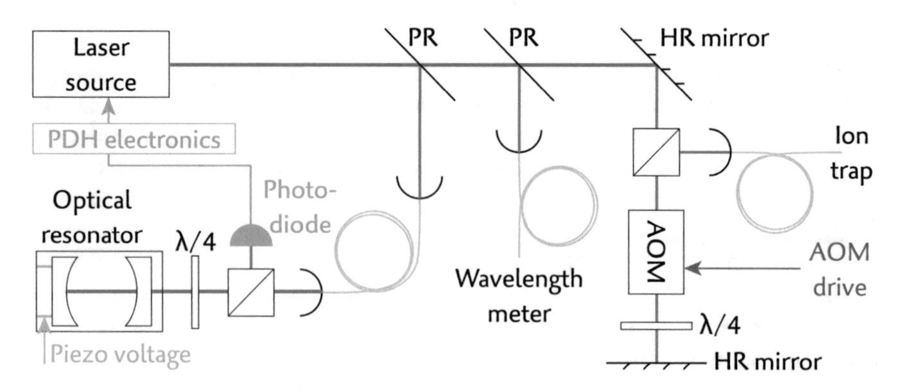

Fig. 3.8 Schematic of a typical laser setup. Highly-reflective and partially-reflective mirrors are labelled HR and PR. Wavelength meter HighFinesse WS6-200 is used

3.3.1 Laser Sources

Laser light for driving the transitions between low-lying states is common to other $^{88}Sr^+$ experiments which study quantum computing [19–21] and atomic clocks [22, 23]. UV laser light for Rydberg excitation is specific to our experiment.

Laser Light for Transitions Between Low-Lying States

We use diode lasers[5] to produce 422, 1033 and 1092 nm laser light and a diode-laser-pumped tapered-amplifier system to produce 674 nm laser light.[6]

Laser Light for the First Rydberg-Excitation Step

Laser light at 243 nm for the first Rydberg-excitation step is produced in a commercial system,[7] as follows: Laser light at 970 nm is produced by a diode-laser-pumped tapered-amplifier system. Two cascaded second harmonic generation (SHG) stages are used to upconvert the 970 nm fundamental to 243 nm laser light. Each SHG stage consists of a nonlinear crystal in a bow-tie cavity. Crystals of lithium triborate (LBO) and barium metaborate (BBO) are used in the first and second stages respectively. Similar laser systems are used for spectroscopy of atomic hydrogen [24].

Laser Light for the Second Rydberg-Excitation Step

Tunable laser light between 304 and 309 nm is required for the second Rydberg-excitation step. The tunability allows a range of states to be excited, from principal quantum number $n = 24$ up to the second ionisation threshold. This laser light is produced in two steps:

First 1551 nm laser light from a diode-laser[8] fibre-amplifier[9] system and tunable laser light between 1000 and 1030 nm from a diode-laser-pumped tapered-amplifier system[10] pass through a periodically-poled lithium niobate (PPLN) crystal[11] and tunable laser light in the range 608–618 nm is produced by sum frequency generation

[5]Toptica DL pro.

[6]Toptica TA pro; a high-power source is used because the 674 nm $5S_{1/2} \leftrightarrow 4D_{5/2}$ transition is electric-dipole-forbidden and relatively high laser light intensities are required to drive it.

[7]Toptica TA-FHG pro.

[8]Originally a TeraXion PS-NLL DFB semiconductor laser was used. The power spectrum consists of a Lorentzian line superposed onto a broad envelope. While 80% of the laser power lies within $2\pi \times 12$ kHz of the centre frequency, 8% of the laser power lies outside $2\pi \times 1$ MHz of the centre frequency (using an observation time of 1 ms). The broad envelope does not allow stable locking of the 608–618 nm laser light to the reference cavity, and it inhibits coherent excitation of Rydberg states using 304–309 nm laser light. We now use a NKT Koheras BASIK E15 DFB fibre laser which has a narrow power spectrum and allows stable locking of the 608–618 nm laser light.

[9]Manlight EYFA-CW-SLM-P-TKS.

[10]Toptica TA pro.

[11]Covesion MSFG612-0.5-40.

(SFG). To achieve quasi-phase matching across the entire 608–618 nm range the crystal temperature is varied and crystal channels with different poling periodicities are used. This first step was the master's thesis work of Christine Maier [25].

In the second step 608–618 nm laser light is upconverted to 304–309 nm laser light in a commercial system consisting of a LBO crystal in a bow-tie cavity,[12] here angle-phase matching is used. Similar laser systems are used in experiments with trapped $^{9}Be^{+}$ ions [26, 27].

Laser Light for Photoionisation of ^{88}Sr Atoms

Ion loading (Sect. 3.2.7) involves two-step photoionisation of neutral ^{88}Sr atoms using laser light produced by diode lasers at 405[13] and 461 nm.[14]

3.3.2 Laser Pulse Generation Using AOMs

Most of the lasers setups include AOMs in double-pass configurations [28]. Laser pulses are engineered by controlling the amplitude, frequency and phase of radiofrequency AOM drive signals (Sect. 3.4). Laser pulses can be produced with constant phase relations between them. The time taken for sound waves in an AOM crystal to move across a laser beam limits the minimum pulse length to ~50 ns.

3.3.3 Optical Fibres

Single-mode fibres allow stable beam pointing onto the ions and they also clean up the laser beams. Polarisation-maintaining single-mode step-index fibres transmit 405, 422, 461, 674, 1033 and 1092 nm laser light close to the experiment chamber. Before focussing onto the ions 422, 1033 and 1092 nm laser light is overlapped in an endlessly single-mode photonic crystal fibre.

Single-Mode Fibres for UV Laser Light

Conventional single-mode step-index fibres made of silicon dioxide are unsuitable for transmitting UV laser light; they degrade due to ultraviolet solarisation. We use hydrogen-loaded photonic crystal fibres with relatively large mode field diameters

---

[12]Toptica SHG pro.

[13]InsaneWare blu-ray diode used originally, recently replaced by a fibre-coupled system Thorlabs LP405-SF10.

[14]Toptica DL pro.

$\sim$8 μm which resist UV solarisation [29].[15] Details can be found in the master's thesis of Johannes Haag [30].

Typically up to 3 mW of 243 nm laser light and 50 mW of 304–309 nm laser light is focussed onto the ions. Techniques developed for focussing these laser beams onto trapped ions are described in Sect. 4.2.

3.3.4 Laser Frequency Stabilisation

To efficiently drive the transitions in Fig. 3.1 lasers must be resonant with the transitions and their linewidths must be lower than the corresponding transition linewidths. Most of the diode lasers have free-running linewidths $\sim 2\pi \times 200$ kHz. The lasers' linewidths are reduced and their centre frequencies are controlled by using optical resonators as frequency references. The Pound–Drever–Hall (PDH) technique [31, 32] is used to confer the stability of the lengths of optical resonators onto the frequency stability of lasers.

Most of the lasers are stabilised using optical resonators built in-house. The 674 nm laser is locked to a highly-stable commercial optical resonator which has a high finesse.

Stabilisation Using Optical Resonators Built in-House

Eight optical resonators with stable lengths were built following the design in [33]. A clear schematic is in [34].

In each optical resonator setup one of the mirrors is mounted on piezoelectric rings; by changing the voltages applied to the piezo rings the length of each resonator is controlled. Each resonator is used to stabilise the frequency of one laser. Because the length of each resonator can be changed by more than half of the laser wavelength, the stabilised lasers can be tuned to any frequency.

The eight optical resonators were designed for eight different wavelengths of laser light: 422, 608–618, 674, 970, 1000–1030, 1033, 1092 and 1550 nm. The mirror coatings of each of the resonators were chosen for a particular laser wavelength.

The 422, 1033 and 1092 nm lasers are stabilised to optical resonators with finesses $\sim$1000 and the stabilised laser linewidths are $\sim 2\pi \times 100$ kHz. The 608–618 and 970 nm laser light is stabilised to optical resonators with finesses $\sim$15000. The 608–618 nm laser light is locked by sending feedback to the 1000–1030 nm fundamental laser. The frequency stability of the fundamental lasers is conferred onto the Rydberg-excitation laser light; this results in laser light with linewidth $\sim 2\pi \times 100$ kHz at both 243 and 304–309 nm.

Optical resonators for 1000–1030 and 1550 nm laser light are used for diagnostic purposes. An optical resonator for 674 nm laser light was set up. It has finesse $\sim$10000

[15]NKT LMA-10-UV.

and it may be used in the future to implement a two-stage laser lock [34, 35] together with the commercial high-finesse optical resonator described below.

The stabilised laser linewidths are estimated by an in-loop method: the gradient of the PDH error signal is used to convert the standard deviation of the error signal of a locked laser to a laser linewidth.

Sidebands are required for laser frequency stabilisation by the PDH technique. Sidebands are introduced by modulating the laser diode current of the 422, 970, 1000–1030, 1033 and 1092 nm lasers. The sidebands persist in the laser light produced by SFG and SHG, and they are used to stabilise the lengths of the SHG bow-tie cavities to the laser frequencies by employing the PDH technique.

Stabilisation Using the High-Finesse Resonator

For coherent operations between the qubit states (sublevels of $5S_{1/2}$ and $4D_{5/2}$) it is necessary to have a highly-stable 674 nm laser. Using a commercial optical resonator[16] with a high-finesse ~100000 the laser frequency was stabilised to $\sim 2\pi \times 1$ kHz during the master's thesis work of Florian Kress [36].

Because sidebands in the laser light would disrupt coherent manipulation of the ion, we introduce sidebands to the 674 nm laser using an electro-optic modulator immediately before the optical resonator. The length of the high-finesse cavity cannot be scanned; to make up for the $\approx 2\pi \times 1.41$ GHz mismatch between the high-finesse cavity resonance and the $5S_{1/2} \leftrightarrow 4D_{5/2}$ atomic resonance multiple AOMs are used.

Frequency drifts of the high-finesse resonator are countered by using the ion as a frequency reference in a Ramsey-type scheme described in [37, 38]. This method also allows fluctuations in the magnetic field strength at the position of the ion to be compensated such that transitions between different Zeeman sublevels of $5S_{1/2}$ and $4D_{5/2}$ may be resonantly driven.

3.4 Electronics for Experimental Control

A PC controls two electronic systems which are used for running experiments: the pulse sequencer box and a bus system. The pulse sequencer box controls the timing of each experiment run. It outputs digital signals and RF signals with a timing resolution of 10 ns. It is described in detail in [39, 40]. The bus system also controls RF signals, though its outputs are updated over microsecond timescales. It is described in detail in [34, 41]. A schematic representation of the control systems is in Fig. 3.9. We use the Trapped Ion Control Software (TrICS) developed by the Blatt group in Innsbruck as a user interface for controlling these systems and collecting experimental data.

During a single experiment run a sequence of laser pulses is applied to trapped ions and then the ions' states are measured (Sect. 3.1). The laser pulses are produced by driving AOMs with RF signals (Sect. 3.3.2). RF signals with 10 ns timing resolution

[16] Stable Laser Systems.

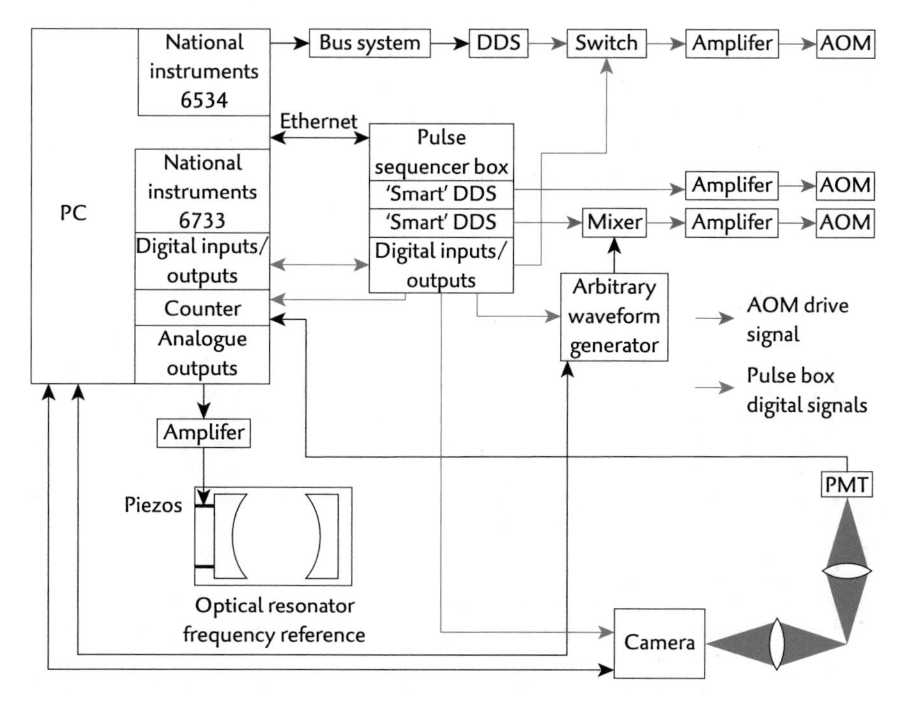

Fig. 3.9 Schematic of electronic control systems

are output directly from the pulse sequencer box or they are produced by switching the RF signals from the bus system using digital signals from the pulse sequencer box. Measurements are choreographed by using digital signals from the pulse sequencer box to gate the photon counter and to trigger the EMCCD camera.

The radiofrequency AOM drive signals are produced by DDS chips. The DDS chips are either programmed via the bus system or they are included in the pulse sequencer box. The "smart" DDS boards within the pulse sequencer box produce versatile AOM drive signals which allow phase-coherent frequency switching and amplitude shaping of laser pulses during a single experiment run. Phase-coherent frequency switching is required for coherent manipulation of an ion state involving different transitions [40]. The amplitude shaping has bandwidth $\approx 2\pi \times 5\,\text{MHz}$. The "smart" DDSs are usually reserved for control of the 674 nm laser beams and the Rydberg-excitation lasers.

Digital signals from the pulse sequencer box are used to trigger and switch other devices, such as arbitrary waveform generators used to shape laser pulses with high bandwidth ($\sim 2\pi \times 20\,\text{MHz}$) for coherent Rydberg excitations (Chap. 7).

Experiment runs are typically repeated ~ 100 times with the same experimental parameters, and sets of experiment runs are repeated as a parameter is scanned. Laser frequencies, powers and phases are scanned by reprogramming the bus system DDSs or the pulse sequencer box settings between sets of experiment runs.

The PC communicates with the bus system via National Instruments card 6534 and it communicates with the pulse sequencer box via National Instruments card 6733 and an Ethernet connection. Card 6733 includes the counter which records signals from the PMT (see Sect. 3.1.4) and it supplies analogue voltages which control the lengths of the optical resonators used as frequency references (Sect. 3.3.4).

References

1. Biémont E et al (2000) Lifetimes of metastable states in Sr II. Eur Phys J D 11:355–365
2. Safronova UI (2010) All-order perturbation calculation of energies, hyperfine constants, multipole polarizabilities, and blackbody radiation shift in $^{87}Sr^{+}$. Phys Rev A 82:022504
3. Letchumanan V, Wilson MA, Gill P, Sinclair AG (2005) Lifetime measurement of the metastable $4d2^{D}_{5/2}$ state in $^{88}Sr^{+}$ using a single trapped ion. Phys Rev A 72:012509
4. Pinnington EH, Berends RW, Lumsden M (1995) Studies of laser-induced fluorescence in fast beams of Sr+ and Ba+ ions. J Phys B 28:2095
5. Zhang H et al (2016) Iterative precision measurement of branching ratios applied to 5P states in 88Sr+. New J Phys 18:123021
6. Roos C (2000) Controlling the quantum state of trapped ions. PhD thesis, Universität Innsbruck
7. Pokorny F (2014) Experimental setup for trapping strontium Rydberg ions. Master's thesis, Universität Innsbruck
8. Guggemos M (2017) Precision spectroscopy with trapped $^{40}Ca^{+}$ and $^{27}Al^{+}$ions. PhD thesis, Universität Innsbruck
9. Chwalla M (2009) Precision spectroscopy with $^{40}Ca^{+}$ ions in a Paul trap. PhD thesis, Universität Innsbruck
10. Michaelson HB (1977) The work function of the elements and its periodicity. J Appl Phys 48:4729–4733
11. Feldker T (2017) Rydberg excitation of trapped ions. PhD thesis, Johannes Gutenberg-Universität Mainz
12. Feldker T et al (2015) Rydberg excitation of a single trapped ion. Phys Rev Lett 115:173001
13. Brownnutt M, Kumph M, Rabl P, Blatt R (2015) Ion-trap measurements of electric-field noise near surfaces. Rev Mod Phys 87:1419–1482
14. Niedermayr M (2015) Cryogenic surface ion traps. Ph.D thesis, Universität Innsbruck
15. Brownnutt M et al (2007) Controlled photoionization loading of 88Sr+ for precision ion-trap experiments. Appl Phys B 87:411–415
16. Berkeland DJ, Miller JD, Bergquist JC, Itano WM, Wineland DJ (1998) Minimization of ion micromotion in a Paul trap. J Appl Phys 83:5025–5033
17. Keller J, Partner HL, Burgermeister T, Mehlstäubler TE (2015) Precise determination of micromotion for trapped-ion optical clocks. J Appl Phys 118:104501
18. MüCller M, Liang L, Lesanovsky I, Zoller P (2008) Trapped Rydberg ions: from spin chains to fast quantum gates. New J Phys 10:093009
19. Brandl MF et al (2016) Cryogenic setup for trapped ion quantum computing. Rev Sci Instrum 87:113103
20. Mehta KK et al (2016) Integrated optical addressing of an ion qubit. Nat Nanotechnol 11:1066
21. Manovitz T et al (2017) Fast dynamical decoupling of the Mølmer- Sørensen entangling gate. Phys Rev Lett 119:220505
22. Barwood GP et al (2014) Agreement between two 88Sr+ optical clocks to 4 parts in 1017. Phys Rev A 89:050501
23. Dubé P, Bernard JE, Gertsvolf M (2017) Absolute frequency measurement of the 88Sr+ clock transition using a GPS link to the SI second. Metrologia 54:290

24. Kolachevsky N, Alnis J, Bergeson SD, Hänsch TW (2006) Compact solid-state laser source for 1S–2S spectroscopy in atomic hydrogen. Phys Rev A 73:021801
25. Maier C (2013) Laser system for the Rydberg excitation of strontium ions. Master's thesis, Universität Innsbruck
26. Wilson AC et al (2011) A 750-mW, continuous-wave, solid-state laser source at 313nm for cooling and manipulating trapped 9Be+ ions. Appl Phys B 105:741–748
27. Lo H-Y et al (2014) All-solid-state continuous-wave laser systems for ionization, cooling and quantum state manipulation of beryllium ions. Appl Phys B 114:17–25
28. Donley EA, Heavner TP, Levi F, Tataw MO, Jefferts SR (2005) Double-pass acousto-optic modulator system. Rev Sci Instrum 76:063112
29. Colombe Y, Slichter DH, Wilson AC, Leibfried D, Wineland DJ (2014) Single-mode optical fiber for high-power, low-loss UV transmission. Opt Express 22:19783
30. Haag J (2015) Glasfasern als Wellenleiter für ultraviolettes Licht. Master's thesis, Universität Innsbruck
31. Drever RWP et al (1983) Laser phase and frequency stabilization using an optical resonator. Appl Phys B 31:97–105
32. Black ED (2001) An introduction to Pound-Drever-Hall laser frequency stabilization. Am J Phys 69:79–87
33. Kumph M (2015) 2D arrays of ion traps for large scale integration of quantum information processors. PhD thesis, Universität Innsbruck
34. Brandl M (2017) Towards cryogenic scalable quantum computing with trapped ions. PhD thesis, Universität Innsbruck
35. Ludlow AD et al (2007) Compact, thermal-noise-limited optical cavity for diode laser stabilization at 1×10^{-15}. Opt Lett 32:641–643
36. Kress F (2015) Frequenzstabilisierung eines 674 nm Diodenlasers zur Detektion der Rydberganregung von Strontiumionen. Master's thesis, Universität Innsbruck
37. Benhelm J (2008) Precision spectroscopy and quantum information processing with trapped calcium ions. PhD thesis, Universität Innsbruck
38. Kirchmair G (2010) Quantum non-demolition measurements and quantum simulation. PhD thesis, Universität Innsbruck
39. Pham PTT (2005) A general-purpose pulse sequencer for quantum computing. Master's thesis, Massachusetts Institute of Technology
40. Schindler P (2008) Frequency synthesis and pulse shaping for quantum information processing with trapped ions. Master's thesis, Universität Innsbruck
41. Schreck F (2015) Control system. http://www.strontiumbec.com/. Accessed 20 Nov 2017

Chapter 4
Two-Photon Rydberg Excitation

This chapter is concerned with the first Rydberg ion experiments carried out in our laboratory. We use a two-photon excitation scheme, which differs notably from the single-photon excitation scheme used in the experiment in Mainz, described in Sect. 1.3.2. The two-photon excitation scheme allows for a lower light-induced coupling between internal and external degrees of freedom compared with the single-photon excitation scheme (Sect. 4.3). This allows us to observe narrower Rydberg resonances than the Mainz experiment and to resolve resonance structure.

The experimental techniques developed in this chapter allow us to investigate Rydberg ion-trap effects (Chap. 6) and to coherently control Rydberg ions (Chap. 7).

4.1 The First Excitation Step

In our experiment a trapped $^{88}\mathrm{Sr}^+$ ion is excited to Rydberg S- and D-states by two-photon excitation, driven by 243 nm and ~307 nm laser light (see Fig. 3.1). 243 nm laser light drives the first excitation step from the metastable state $4D_{5/2}$ near to $6P_{3/2}$ or from the metastable state $4D_{3/2}$ near to $6P_{1/2}$.

4.1.1 $4D_{5/2} \leftrightarrow 6P_{3/2}$ Transition

The $4D_{5/2} \leftrightarrow 6P_{3/2}$ transition is probed as follows [schematic in Fig. 4.1a]:

0. The ion is prepared in a Zeeman sublevel of $4D_{5/2}$, as described in Sect. 3.1.3.
1. The $4D_{5/2} \leftrightarrow 6P_{3/2}$ transition is driven by 243 nm laser light; excited population decays mostly to $5S_{1/2}$.

G. Higgins, *A Single Trapped Rydberg Ion*, Springer Theses,
https://doi.org/10.1007/978-3-030-33770-4_4

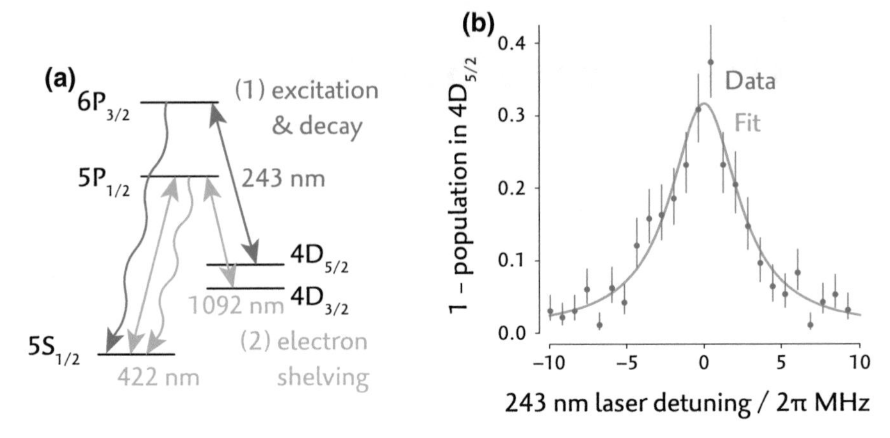

Fig. 4.1 Measurement of the $4D_{5/2} \leftrightarrow 6P_{3/2}$ resonance. **a** First 243 nm light drives the transition to $6P_{3/2}$ and population in $6P_{3/2}$ decays to $5S_{1/2}$, the ion state is then measured by electron shelving. **b** Because detection of successful excitation involves absorption of a 243 nm photon and optical pumping out of $4D_{5/2}$, the resonance lineshape is the Lorentzian absorption profile in the exponent of the exponential function. Error bars indicate quantum projection noise (68% confidence interval)

2. Electron shelving: Detection of fluorescence from the $5S_{1/2} \leftrightarrow 5P_{1/2}$ transition heralds successful excitation to $6P_{3/2}$ (see Sect. 3.1.4).

Using calculated electronic wavefunctions (Sect. 2.2) we find population in $6P_{3/2}$ decays mostly to $5S_{1/2}$ by multi-step decay processes. The resonance lineshape is shown in Fig. 4.1b.

The excitation proceeds incoherently because the transition Rabi frequency is much lower than the $6P_{3/2}$ decay rate. The entire process can be viewed as optical pumping from $4D_{5/2}$ to $5S_{1/2}$ and to $4D_{3/2}$ via $6P_{3/2}$, and the population in $4D_{5/2}$ decays according to

$$P_{4D_{5/2}} = e^{-\mathcal{R}(\omega_{243})t_{\mathrm{ex}}}, \tag{4.1}$$

where the frequency-dependent absorption rate

$$\mathcal{R}(\omega_{243}) = \frac{\Omega_1^2}{\Gamma_{6P_{3/2}}}\left(1 + \frac{4(\omega_{243} - \omega_0)^2}{\Gamma_{6P_{3/2}}^2}\right)^{-1} \tag{4.2}$$

and t_{ex} is the excitation time, Ω_1 is the transition Rabi frequency, $\Gamma_{6P_{3/2}}$ is the natural decay rate of $6P_{3/2}$, ω_{243} is the 243 nm laser frequency and ω_0 is the resonance frequency of the transition. $\mathcal{R}(\omega_{243})$ has a Lorentzian profile with linewidth $\Gamma_{6P_{3/2}}$ unless there is Doppler broadening. From the fit in Fig. 4.1b we extract $\Gamma_{6P_{3/2}} = 2\pi \times (4.9 \pm 0.4)$ MHz, the theoretically-determined values are $2\pi \times 4.71$ MHz [1] and $2\pi \times 4.26$ MHz [2]. We also extract Ω_1 and ω_0 from the fit.

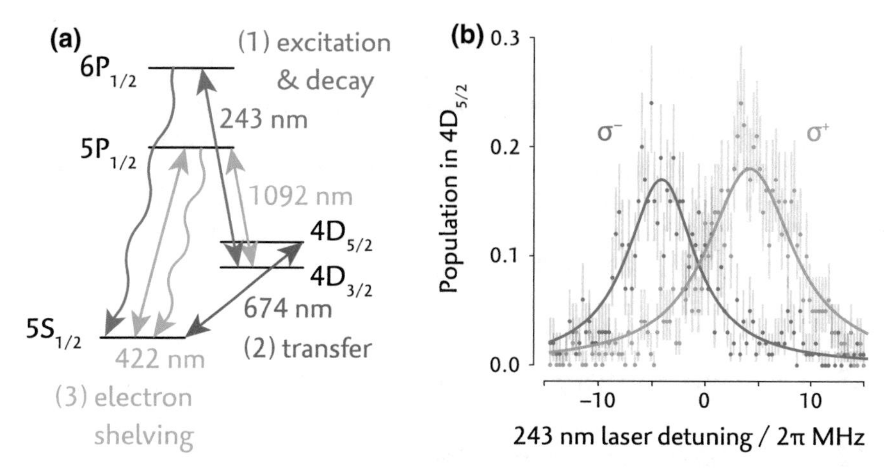

Fig. 4.2 Measurement of the $4D_{3/2} \leftrightarrow 6P_{1/2}$ resonance. **a** First 243 nm light drives the transition to $6P_{1/2}$ and population in $6P_{1/2}$ decays to $5S_{1/2}$, then 674 nm light transfers population from $5S_{1/2}$ to $4D_{5/2}$, finally the ion state is measured. **b** Either σ^+ or σ^- transitions are driven by using the appropriate laser polarisation. The transitions are non-degenerate due to the Zeeman splittings of the atomic sublevels. The solid curves are fits to the experimental data. Error bars indicate quantum projection noise (68% confidence interval)

4.1.2 $4D_{3/2} \leftrightarrow 6P_{1/2}$ Transition

The $4D_{3/2} \rightarrow 6P_{1/2}$ transition is probed as follows [schematic in Fig. 4.2a]:

0. The ion is prepared in a mixture of the four Zeeman sublevels of $4D_{3/2}$ (Sect. 3.1.3).
1. Laser light at 243 nm drives the $4D_{3/2} \leftrightarrow 6P_{1/2}$ transition, population excited to $6P_{1/2}$ decays mostly to $5S_{1/2}$.
2. Population in $5S_{1/2}$ is transferred to $4D_{5/2}$.
3. Electron shelving: Successful excitation to $6P_{1/2}$ is heralded when no fluorescence from the $5S_{1/2} \leftrightarrow 5P_{1/2}$ transition is detected (Sect. 3.1.4).

Using calculated electronic wavefunctions (Sect. 2.2) we find population in $6P_{1/2}$ decays mostly to $5S_{1/2}$ by multi-step decay processes. Resonance lineshapes are shown in Fig. 4.2b.

At the position of the ion the 243 nm laser beam propagation is parallel with the magnetic field and thus laser light with circular polarisation drives the two $4D_{3/2},\ m_J = -\frac{3}{2} \rightarrow 6P_{1/2},\ m_J = -\frac{1}{2}$ and $4D_{3/2},\ m_J = -\frac{1}{2} \rightarrow 6P_{1/2},\ m_J = \frac{1}{2}$ σ^+ transitions or the two respective σ^- transitions.

Each peak in Fig. 4.2b corresponds to two overlapping resonances which are not individually resolved because the $2\pi \times 670$ kHz-Zeeman splitting of the resonance frequencies (in a magnetic field of strength 0.36 mT [1]) is much less than the theory value of the resonance linewidth $\Gamma_{6P_{1/2}} \approx 2\pi \times 4$ MHz [1, 2]. Each peak in Fig. 4.2b

[1]Throughout this work Landé g-factors are used for magnetic moments.

is fit using a similar method as in Sect. 4.1.1, only the absorption rate $\mathcal{R}(\omega_{243})$ includes two Lorentzian functions for the two transitions weighted by their respective Clebsch–Gordan coefficients. The fit function accounts for the imperfect population transfer from $5S_{1/2} \rightarrow 4D_{5/2}$ and assumes the initial population is uniformly spread across the $4D_{3/2}$ sublevels. The frequency difference of the two peaks is consistent with the expected Zeeman splitting between the pairs of transitions. The resonance linewidth inferred from the data $2\pi \times (7.2 \pm 0.4)$ MHz exceeds the theory value $\approx 2\pi \times 4$ MHz [1, 2], possibly due to Doppler broadening.

4.2 Focussing Rydberg Excitation Lasers on an Ion

4.2.1 UV Laser Setups at the Experiment Chamber

Rydberg excitation laser light is delivered near to the experiment chamber through single-mode fibres, as described in Sect. 3.3.3.

After each fibre each laser beam propagates through two identical achromatic lenses and into the experiment chamber. The laser beams propagate through holes in the endcap electrodes [Fig. 3.3a]. The lenses image the fibre modes (with diameters $\sim 8\,\mu$m) onto trapped ions with magnification 1. Because the lenses are achromatic the laser beams may be sent from either direction, or both may be sent from the same direction. A single ion is used to optimise the focussing of each Rydberg excitation laser as described in the following subsections.

4.2.2 Focussing 243 nm *Light on an Ion*

The focussing of 243 nm light onto the ion is optimised by moving the two lenses (which are mounted on translation stages between the fibre output and the experiment chamber) and maximising the strength with which the $4D \leftrightarrow 6P$ transitions in Sect. 4.1 are driven.

The ion is also used to profile the laser beam: The strength with which the 243 nm laser drives the $4D \leftrightarrow 6P$ transitions is recorded as the laser beam is shifted. The laser beam is shifted by displacing the second lens in the directions perpendicular to the laser beam propagation direction. Because the laser beam is collimated at the position of the second lens, displacement of this lens displaces the focus by the same amount at the position of the ion. A beam profile measured using the ion is shown in Fig. 4.3.

At the ion position the beam is elliptical, with Gaussian beam waists $w_x = 4.9\,\mu$m and $w_y = 10.7\,\mu$m. The laser beam propagates through holes in the trap endcaps which limit the numerical aperture to 0.058. In principle a focus with waist $1.3\,\mu$m may be attained with an improved optical setup.

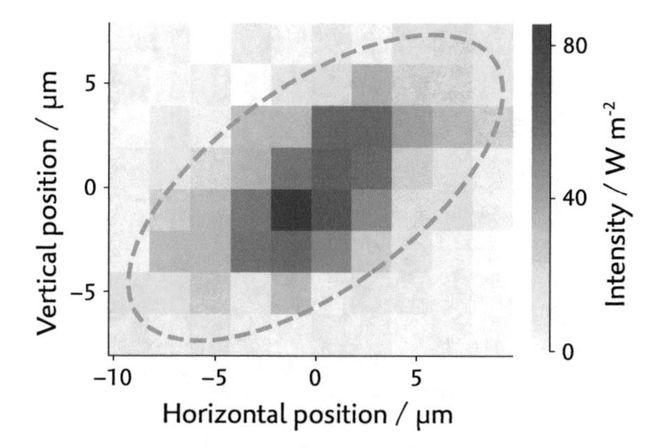

Fig. 4.3 The 243 nm laser beam profile at the position of the ion is determined by moving the laser beam and measuring the strength with which the $4D_{5/2} \leftrightarrow 6P_{3/2}$ transition is driven. The dashed green line represents the $1/e^2$ beam waist. The relative intensity is converted to the absolute laser light intensity with 44% uncertainty. This uncertainty arises because the laser power is not measured at the ion position; rather the laser power is measured before and after the chamber and the surrounding optics, and only 32% of the laser power $[(1-0.44)^2]$ is transmitted through the system

4.2.3 *Focussing* ~307 nm *Light on an Ion*

Laser light near 307 nm is required for the second Rydberg excitation step. This laser light shifts the energies of $5S_{1/2}$ and $4D_{5/2}$ Zeeman sublevels according to the ac-Stark effect, by an amount which is proportional to the 307 nm laser light intensity. The focussing of ~307 nm light onto the ion is optimised by moving the two lenses and maximising the resonance frequency shift induced onto one of the $5S_{1/2} \leftrightarrow 4D_{5/2}$ transitions. The resonance frequency shift is measured in a Ramsey experiment.

This Ramsey-type method is also used to measure the 307 nm laser beam profile at the position of the ion, which is shown in Fig. 4.4. The beam is near-round, with Gaussian beam waists $w_x = 5.3\,\mu$m and $w_y = 6.1\,\mu$m. With the current setup the focus likely cannot be improved, given the expected fibre mode field diameter $\approx 8\,\mu$m.

This laser beam also propagates through holes in the trap endcaps which limit the numerical aperture to 0.058, and thus a beam waist of $1.7\,\mu$m can be attained for this laser beam with an improved optical setup.

4.3 Two-Photon Resonance Condition

Rydberg states are excited and detected using a similar methodology as in Sect. 4.1; UV laser light at 243 nm and ~307 nm drives a two-photon transitions from a metastable $4D$-state to a Rydberg S- or D-state. Each UV laser is detuned from

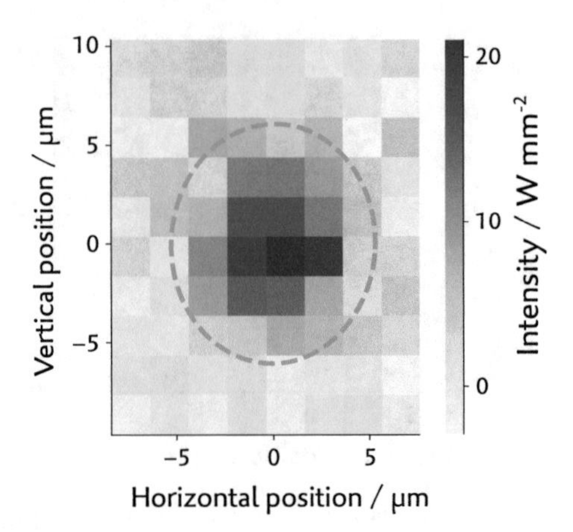

Fig. 4.4 The 307 nm laser beam profile at the position of the ion is determined by moving the laser beam and at each position the relative 307 nm laser light intensity is measured in a Ramsey-type experiment. The dashed green line represents the $1/e^2$ beam waist. The relative intensity is converted to the absolute laser light intensity with 16% uncertainty. This uncertainty arises because the laser power is not measured at the ion position; rather the laser power is measured before and after the chamber and the surrounding optics, and 70% of the laser power $[(1-0.16)^2]$ is transmitted through the system. The intensity scale includes non-physical negative values, this is due to statistical fluctuations of the Ramsey measurement results

resonance to the intermediate $6P$ state by typically $2\pi \times 200$ MHz, the detunings are opposite such that two-photon resonance is achieved. Using calculated electronic wavefunctions (Sect. 2.2) we find population in Rydberg S- and D-states decays mostly to $5S_{1/2}$ by multi-step decay processes.

The effective Rabi frequency of the two-photon transition is

$$\Omega_{\text{eff}} = \frac{\Omega_1 \Omega_2}{2\Delta}, \tag{4.3}$$

where Δ is the magnitude of the detuning of each laser from the intermediate state. This formula is derived in Sect. 7.3.1.

We use $\Delta \gg \Omega_1,\ \Gamma_{6P}$ to minimise scattering off the intermediate state which causes a background signal. The rate at which population from the initial state scatters off the intermediate state is

$$\Gamma_{\text{scat}} = \frac{\Omega_1^2 \Gamma_{6P}}{4\Delta^2}. \tag{4.4}$$

Results are shown in Fig. 4.5 for the excitation from $4D_{3/2} \rightarrow 25S_{1/2}$ which exemplify the two-photon resonance condition. In each dataset the frequency of the 243 nm laser has a different intermediate-state detuning while the frequency of the 309 nm laser is scanned; the Rydberg resonance structure is observed when the two-photon resonance condition is met. Each dataset shows a four-peak resonance structure corresponding to four Zeeman-split transitions, this structure is explained in Sect. 4.5.

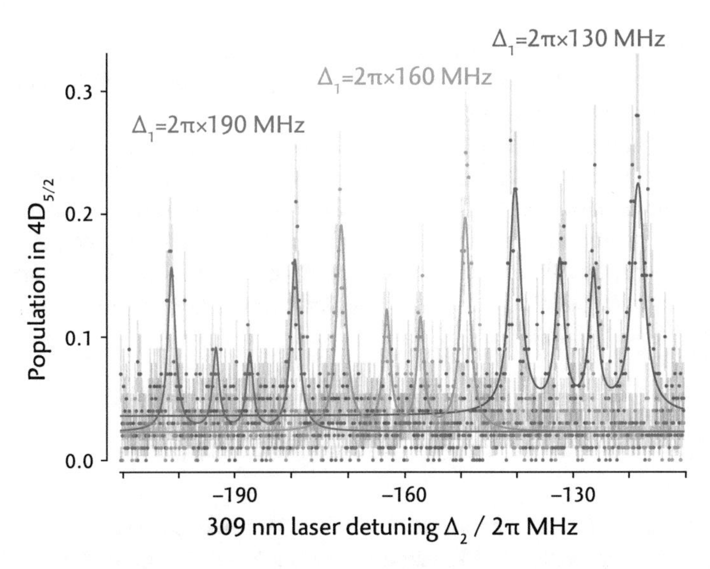

Fig. 4.5 Two-photon excitation from $4D_{3/2}$ to $25S_{1/2}$. The resonance structure is centred on the position where the two-photon detuning $\Delta_1 + \Delta_2 = 0$. The solid lines show fits to the data. The absolute value of the detuning of the 309 nm laser from the $6P_{1/2} \leftrightarrow 25S_{1/2}$ resonance is inferred from all three of the data sets. Error bars indicate quantum projection noise (68% confidence interval)

4.4 Counterpropagating Rydberg-Excitation Lasers

The two UV laser beams counterpropagate along the trap axis, collinear with the magnetic field [Fig. 3.3a]. On two-photon absorption the momentum kicks from the two photons on the ion largely cancel and the ion motion is not greatly disturbed. This means phonon-number-changing transitions (Sect. 3.1.1) are less likely to occur with counterpropagating laser beams than with copropagating laser beams or with a single Rydberg excitation step. More thoroughly, it is easier to achieve excitation within the Lamb-Dicke regime when counterpropagating laser beams are used. By adapting Eq. (3.1) we find the effective Lamb-Dicke parameter

$$\eta_{\mathrm{eff}} = (k_1 - k_2)\sqrt{\frac{\hbar}{2M\omega_z}}, \tag{4.5}$$

where k_1, k_2 are the wavevectors of the 243 nm and ~307 nm laser light, M is the $^{88}Sr^+$ mass, and ω_z is the axial trapping frequency. $\eta_{\mathrm{eff}} \approx 0.045$ when typical axial trapping frequencies are used. After Doppler cooling the inequality in Eq. (3.2) is satisfied and the ion is in the Lamb-Dicke regime with regards the two-photon transition using counter-propagating lasers. Working within the Lamb-Dicke regime allows us to investigate Rydberg ion-trap effects (Chap. 6).

If instead the UV laser beams were copropagating, the Lamb-Dicke parameter would be 9 times larger and sub-Doppler cooling would be required to prepare an ion in the Lamb-Dicke regime.

The opportunity to lower the effective Lamb-Dicke parameter by using counter-propagating laser beams is a major advantage of the two-photon excitation scheme. In the Mainz experiment $\sim$122 nm laser light drives the single-photon Rydberg-excitation transition and sub-Doppler cooling is required to prepare an ion in the Lamb-Dicke regime. Phonon-number-changing transitions are driven during Rydberg excitation in their experiment; this contributes to their relatively broad resonances (see Sect. 1.3.2).

4.5 Zeeman-Split Rydberg S-State Resonances

Transitions driven by the trapping electric fields (Sect. 6.2) are negligible for Rydberg S-states up to at least $n = 50$ [3], and thus m_J is a 'good' quantum number both for low-lying states and for the Rydberg S-states investigated in this work. The Zeeman effect describes the ion in a magnetic field; field strengths $\approx$0.3 mT are used.[2]

The resonance structure obtained when a Rydberg S-state is excited is understood in terms of the Zeeman effect: The four resonance peaks in Figs. 4.5 and 4.6a correspond to the four transitions between sublevels of $4D_{3/2}$ and sublevels of $nS_{1/2}$. The transitions are non-degenerate because the sublevels of $4D_{3/2}$ and of $nS_{1/2}$ are Zeeman-split.

The configuration of the laser beams determines the transitions that may be driven [displayed in Fig. 4.6b]: The two Rydberg-excitation laser beams counter-propagate along the trap axis, collinear with the magnetic field. Each laser beam may drive σ^+ transitions, σ^- transitions, or both transitions depending on the laser light polarisation. The different amplitudes of the resonance peaks result from the different Clebsch–Gordan coefficients of the constituent transitions, which are shown in Fig. 4.6b.

A single transition between Zeeman sublevels may be driven by using the appropriate UV laser polarisations, as shown in Fig. 4.6a, or by relying on the transition being frequency-resolved. When using high Rabi frequencies and short pulse lengths for coherent Rydberg excitation (as in Chap. 7) Fourier broadening causes the frequency-resolving power to be reduced and it is then best to select a single transition using laser polarisations.

[2]The Zeeman effect is much smaller than the fine-structure splitting until $n \sim 200$. The paramagnetic Zeeman term dominates the diamagnetic term also until $n \sim 200$.

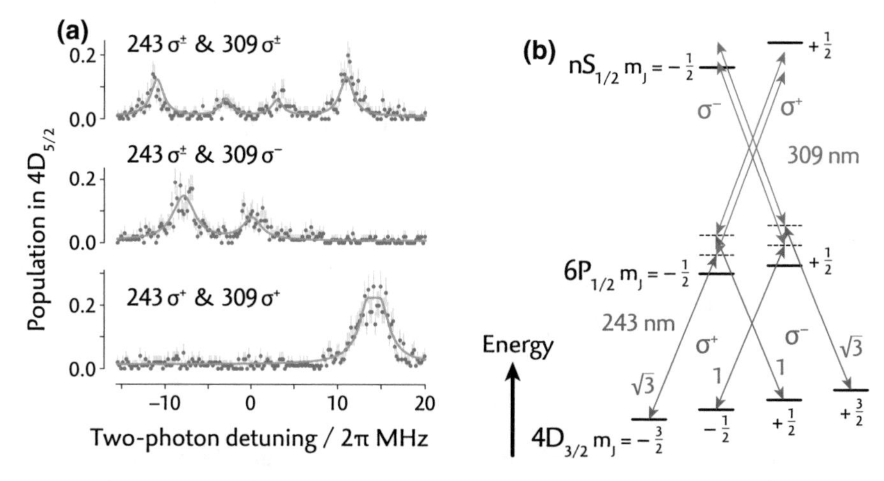

Fig. 4.6 The $4D_{3/2} \leftrightarrow 25S_{1/2}$ resonance structure results from Zeeman splitting of the $4D_{3/2}$ and $25S_{1/2}$ sublevels. At most four non-degenerate two-photon transitions are driven, since the Rydberg excitation laser beams are collinear with the magnetic field. **a** Individual transitions are selected by using appropriate laser polarisations. The solid lines are fits to the data, the magnetic field strength is constrained to 0.356 mT—it is independently determined from the $5S_{1/2} \leftrightarrow 4D_{5/2}$ spectrum. The amplitudes of the peaks depend on the Clebsch–Gordan coefficients of the first transition step; the relative magnitudes of the Clebsch–Gordan coefficients are shown in (**b**). As the 309 nm laser polarisation is changed the resonances shift by $\approx 2\pi \times 2$ MHz likely because of the ac-Stark effect. The fit to the lowest spectrum appears clipped at 0.22 due to saturation of the absorption signal; the maximum signal is limited by $\approx$25% of the population being initialised in each $4D_{3/2}$ sublevel and the $\approx$87% $5S_{1/2} \rightarrow 4D_{5/2}$ transfer efficiency during the measurement step. Error bars indicate quantum projection noise (68% confidence interval)

4.6 Rydberg Energy Series

We have excited a trapped $^{88}\mathrm{Sr}^+$ ion to Rydberg S-states with principal quantum numbers in the range 25–57. The state energies are determined by employing a wavelength meter[3] to measure the frequencies of the fundamental lasers used to generate the Rydberg excitation lasers. Uncertainties in the wavelength measurements limit the determination of the state energies to $\hbar\, 2\pi \times 300$ MHz.

The state energies E_{nLJ} are fit following the methodology of [4] with the Rydberg energy series formula

$$E_{nLJ} = I^{++} - \frac{Z^2 R_M}{(n - \mu(E_{nLJ}))^2}, \tag{4.6}$$

where I^{++} is the second ionisation threshold and the Rydberg constant for $^{88}\mathrm{Sr}^+$ with mass $M = 87.9$ u [5] is

[3] HighFinesse WS6-200.

$$R_M = R_\infty \frac{M}{M + m_e}, \tag{4.7}$$

R_∞ is the Rydberg constant, m_e is the electron mass, and the quantum defect is a linear function of the binding energy

$$\mu(E_{nLJ}) = \mu(I^{++}) + \frac{\partial \mu}{\partial E}(E_{nLJ} - I^{++}) \tag{4.8}$$

$$= \mu(I^{++}) - \frac{\partial \mu}{\partial E} \frac{Z^2 R_M}{(n - \mu(E_{nLJ}))^2} \tag{4.9}$$

$$\approx \mu(I^{++}) - \frac{\partial \mu}{\partial E} \frac{Z^2 R_M}{(n - \mu(I^{++}))^2}. \tag{4.10}$$

The energy series and the mismatch between measured values and the fit are shown in Fig. 4.7. Resonance frequency shifts due to trap effects (Chap. 6) and the ac-Stark effect are small compared with the wavelength meter uncertainty.

The energy series of Rydberg S-, D-, F- and G-states of Sr^+ in free space were previously measured by Lange et al. [4]. They measured states energies accurate to $\hbar\, 2\pi \times 6\,\text{GHz}$, limited by the linewidth of their Rydberg-excitation laser at ~280 nm. Our estimates of the double ionisation threshold and the S-state quantum defect are consistent with the previous experimental investigation, as shown in Table 4.1.

We have also excited two Rydberg D-states and estimate quantum defects $\mu(24D_{3/2}) = 1.4563 \pm 0.0003$ and $\mu(27D_{3/2}) = 1.4563 \pm 0.0004$, which are consistent with the values from Lange et al.

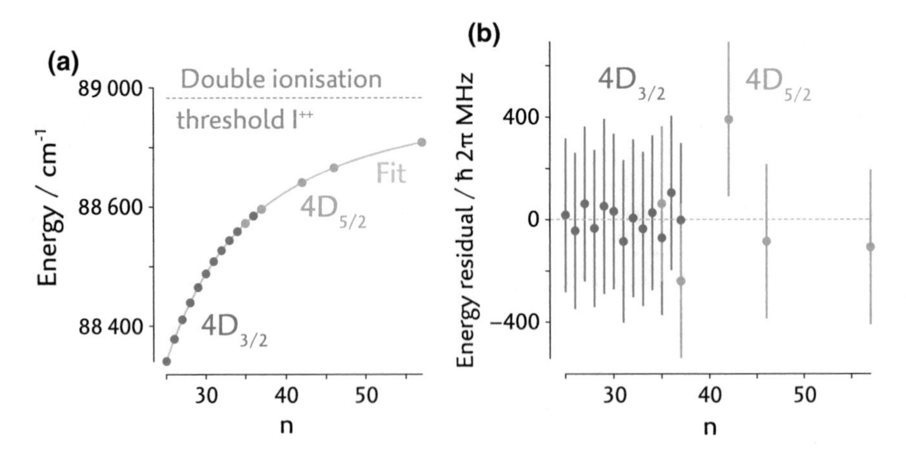

Fig. 4.7 $^{88}Sr^+$ S-state energies are described well by the Rydberg energy series formula. State energies are measured using excitation from $4D_{3/2}$ and from $4D_{5/2}$. **a** Binding energies scale as n^{*-2}. **b** Differences between the data and the fit are generally within the 68% confidence interval indicated by the error bars. Experimental uncertainties are likely correlated; they are dominated by the wavelength meter which contributes $2\pi \times 300\,\text{MHz}$ to the error bars

Table 4.1 The $^{88}Sr^{+}$ double ionisation threshold and S-state quantum defect determined in this work is consistent with previous work [4]. Our value of I^{++} uses the $4D_{5/2}$ level energy from [6]

	This work	Lange et al. [4]
Double ionisation threshold I^{++}/cm^{-1}	88965.022 ± 0.011	88965.18 ± 0.02
S-series quantum defect $\mu(I^{++})$	2.7063 ± 0.0009	2.707 ± 0.002
S-series quantum defect gradient $(\partial\mu/\partial E)\,/\,\mathrm{Ryd}^{-1}$	-0.04 ± 0.09	-0.055 ± 0.015

Improved spectroscopy allows for more accurate calculations of electronic wavefunctions (Sect. 2.2) and thus more accurate theoretical predictions of Rydberg state properties. Our institute has recently purchased a wavelength meter accurate to $2\pi \times 2$ MHz.[4] We plan to repeat the energy series measurement of Rydberg S-states, as well as Rydberg P- and D-states.[5]

References

1. Biémont E et al (2000) Lifetimes of metastable states in Sr II. Eur Phys J D 11:355–365
2. Safronova UI (2010) All-order perturbation calculation of energies, hyperfine constants, multipole polarizabilities, and blackbody radiation shift in $^{87}Sr^{+}$. Phys Rev A 82:022–504
3. Müller M, Liang L, Lesanovsky I, Zoller P (2008) Trapped Rydberg ions: from spin chains to fast quantum gates. New J Phys 10:093009
4. Lange V, Khan MA, Eichmann U, Sandner W (1991) Rydberg states of the strontium ion. Z Phys D 18:319–324
5. Audi G, Bersillon O, Blachot J, Wapstra A (2003) The Nubase evaluation of nuclear and decay properties. Nucl Phys A 729:3–128
6. Barwood GP et al (2014) Agreement between two $^{88}Sr^{+}$ optical clocks to 4 parts in 10^{17}. Phys Rev A 89:050501

[4]HighFinesse WS8-2.

[5]Rydberg P-states were recently excited in our experiment using two UV photons and one MW photon; this work forms part of the PhD thesis project of Fabian Pokorny, see the Outlook in Chap. 8.

Chapter 5
Ion Loss by Double Ionisation

In our experiment a trapped $^{88}\mathrm{Sr}^+$ ion is lost by double ionisation after typically several hundred excitations to a Rydberg state. Similarly double ionisation occurs in $\sim$0.3% excitations to a Rydberg state in the Mainz experiment [1]. Double ionisation thus presents an obstacle in trapped Rydberg ion experiments, discussed in Sect. 5.1. We know the ion is lost by double ionisation by measuring the final product to be $^{88}\mathrm{Sr}^{2+}$, these measurements are described in Sect. 5.2.

We suspect Rydberg states with higher principal quantum numbers are more prone to double ionisation loss and that blackbody radiation increases the likelihood of double ionisation, though we have yet to carry out a systematic investigation of the effects of different parameters on the ion loss rate. Müller et al. [2] briefly discuss double ionisation driven by the electric fields of the trap.

5.1 Obstacles Presented by Ion Loss

Ion loss has made data collection in this experiment cumbersome. Until recently ion loading was carried out using a resistively-heated oven (Sect. 3.2.7) and each double ionisation event caused the experiment to be interrupted by around 20 minutes. During these recurrent interruptions the frequencies of the Rydberg excitation lasers drifted by $\sim 2\pi \times 300\,\mathrm{kHz}$ because the lengths of the optical resonator frequency references (Sect. 3.3.4) are not perfectly stable. Interruptions due to ion loss are now reduced to less than 30 s through introduction of laser ablation loading. A Rydberg ion quantum computer would certainly require faster ablation loading [3, 4] or fast loading from an ion reservoir [1].

Ion loss threatens the gate fidelity in a Rydberg ion quantum computer, though it may be tolerable. Provided the loss rate is below a threshold value, quantum error correction protocols may be used to correct for ion loss. For instance a single ionic qubit lost due to double ionisation may be replaced by shuttling an ion in state $|1\rangle$ from a reservoir [5], the qubit state may then be corrected using the same quantum error correction protocols which correct for qubit state decay $|0\rangle \rightarrow |1\rangle$ [6].

G. Higgins, *A Single Trapped Rydberg Ion*, Springer Theses,
https://doi.org/10.1007/978-3-030-33770-4_5

Atom loss presents similar difficulties in neutral Rydberg atom experiments [7, 8]. In optical traps intense trapping fields drive photoionisation of Rydberg states at rates which may exceed radiative decay rates [9–11]. To get around this problem optical traps are routinely switched off during Rydberg excitation,[1] and some researchers are pursuing blue-detuned (dark) optical traps [8, 12]. Blackbody radiation also drives photoionisation of Rydberg atoms [13]. Additionally ground-state atoms are lost from optical traps due to collisions with background gas particles.

5.2 Measurements of the Loss Product

We know the ion is lost by double ionisation because we measure the loss product to be $^{88}Sr^{2+}$. In Sect. 5.2.1 we establish the loss product has charge +2 e by imaging ions on an electron-multiplying charge-coupled device (EMCCD) camera. In Sect. 5.2.2 we find the loss product has the same mass as $^{88}Sr^{+}$ by measuring motional mode frequencies.

We also find that ions may be doubly-ionised not only from a Rydberg state, but also from a metastable $4D$ state when intense 243 nm laser light is used. In such cases double ionisation likely proceeds via absorption of two 243 nm photons.

5.2.1 *Imaging Ions with an EMCCD Camera*

$^{88}Sr^{+}$ ions scatter 422 nm laser light, and are imaged on an EMCCD camera, as shown in Fig. 5.1. Multiple trapped ions crystallise as a linear string along the trap axis when the confining potential is much weaker in the axial direction than the radial directions. The equilibrium positions of multiple trapped ions are found by balancing the Coulomb repulsion between the ions and the confining harmonic trap forces [14].

Two ions each with charge +e have equilibrium positions $\pm 2^{-2/3}L$ relative to the trap centre, where

$$L = \sqrt[3]{\frac{e^2}{4\pi\epsilon_0 M\omega_z^2}}, \tag{5.1}$$

e is the elementary charge, M is the ion mass and ω_z is the axial mode frequency. ω_z is determined by spectroscopy on the qubit transition or by 'tickling' the trap electrons (see Sect. 5.2.2). By imaging two ions on the EMCCD camera[2] the image magnification is determined to be 19.2.

[1]Switching off the trap is not an option in our experiment because Coulomb repulsion between two ions initially separated by 5 μm would cause them to become separated by 30 μm in just 1 μs.

[2]Andor iXon3 897 with pixel size 16 μm.

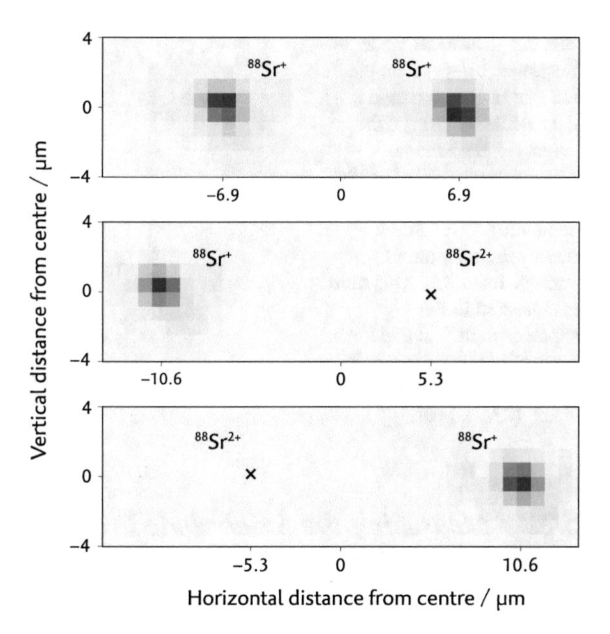

Fig. 5.1 $^{88}Sr^{+}$–$^{88}Sr^{+}$ ion crystal compared with $^{88}Sr^{+}$–$^{88}Sr^{2+}$ crystals. $^{88}Sr^{+}$ ions are imaged on an EMCCD camera using scattered 422 nm laser light, $^{88}Sr^{2+}$ ions are not observed directly. In the lower two images the distance of the $^{88}Sr^{+}$ ion from the trap centre is consistent with a mixed crystal containing a doubly-charged ion. In each image the X marks the spot where the $^{88}Sr^{2+}$ ion lies

The equilibrium positions of ions with different charges is determined by generalising the calculation in [14]. An ion with charge +e and an ion with charge +2 e form crystals with equilibrium positions $\{\pm 2 \times 3^{-2/3}L, \mp 3^{-2/3}L\}$ relative to the trap centre.

We start with two trapped $^{88}Sr^{+}$ ions, shown in Fig. 5.1. We then alter one of the ions during Rydberg excitation. The altered ion remains trapped, although it does not scatter 422 nm laser light. The unmodified $^{88}Sr^{+}$ ion is still visible on the EMCCD camera and may be stably trapped at positions $\pm 2 \times 3^{-2/3}L$, as shown in the lower images of Fig. 5.1. From this we determine the altered ion (which is not imaged) has +2 e charge.

Crystals consisting of one ion with charge +e and one ion with charge +2 e are also obtained when we begin with only one $^{88}Sr^{+}$ ion in the trap and then alter it during Rydberg excitation before loading another $^{88}Sr^{+}$ ion into the trap. This indicates the double ionisation process does not rely upon having two ions in the trap. The same crystal is also obtained when intense 243 nm laser light is used to doubly-ionise an ion and no 307 nm laser light is employed.

The equilibrium ion positions are obtained by balancing electrostatic forces; they do not depend upon ion masses. Equation (5.1) shows no mass dependence, since $\omega^2 \sim \frac{1}{M}$. The mass of the doubly-charged ion is determined by measuring motional mode frequencies, as is described in the following section.

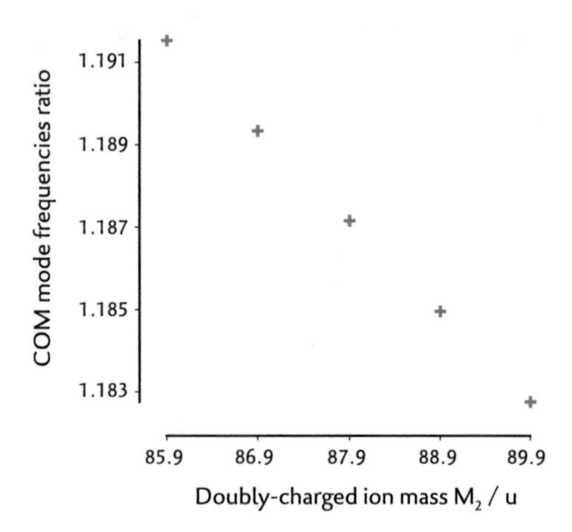

Fig. 5.2 Motional mode frequencies depend on the ion masses. Shown here is the calculated ratio of the centre-of-mass mode frequency for a $^{88}Sr^{+}$–$^{88}Sr^{+}$ crystal and a mixed crystal containing $^{88}Sr^{+}$ and a doubly-charged ion of variable mass M_2. This ratio is measured in the experiment to find that the mass of the loss product is consistent with the mass of $^{88}Sr^{+}$ (87.9 u [16])

5.2.2 *Measuring the Axial Mode Frequency by Tickling Electrodes*

We measure the mass of the altered ion to determine the loss product is $^{88}Sr^{2+}$ rather than some other (possibly molecular) doubly-charged species. We measure the mass of the altered ion with accuracy better than 1 u, and find the mass is consistent with the mass of $^{88}Sr^{+}$. We do so by measuring the frequency of a motional mode shared by a $^{88}Sr^{+}$ ion and the doubly-charged loss product.

The motional modes of N ions of the same mass and charge confined in a harmonic potential and arranged in a linear string is found in [14] by first using the potential of the N-ion system to find the equilibrium positions $\{z_{i,0}\}$, then writing the Lagrangian in terms of small displacements from the equilibrium positions; for the ith ion the displacement $q_i(t) = z_i(t) - z_{i,0}$. Generalising this to a string of ions with different masses and different charges involves writing the Lagrangian in terms of mass-weighted displacements $p_i(t) = \sqrt{M_i} q_i(t)$. The same rescaling is applied in [15] to find the motional modes of different species of ions in an anharmonic trap.

Two ions with the same mass and the same charge trapped in a harmonic trap with frequency ω have motional modes with frequencies ω (the centre-of-mass mode) and $\sqrt{3}\omega$ (the breathing mode). If the mass and charge of one of the ions is different, the motional mode frequencies change. We calculate the centre-of-mass mode frequencies for one $^{88}Sr^{+}$ ion (charge +e, mass 87.9 u [16]) trapped together with an ion with charge +2 e and mass M_2; results are shown in Fig. 5.2. A change in M_2 of 1 u results in a change in the mixed-species centre-of-mass mode frequency of around $2 \times 10^{-3} \times \omega$. In the experiment we measure the centre-of-mass frequencies of both the $^{88}Sr^{+}$–$^{88}Sr^{+}$ crystal and the mixed-species crystal. We determine the ratio between these frequencies with accuracy $<2 \times 10^{-4}$ and find no significant difference between the loss product mass M_2 and the $^{88}Sr^{+}$ mass 87.9 u. In this way we determine the loss product is $^{88}Sr^{2+}$.

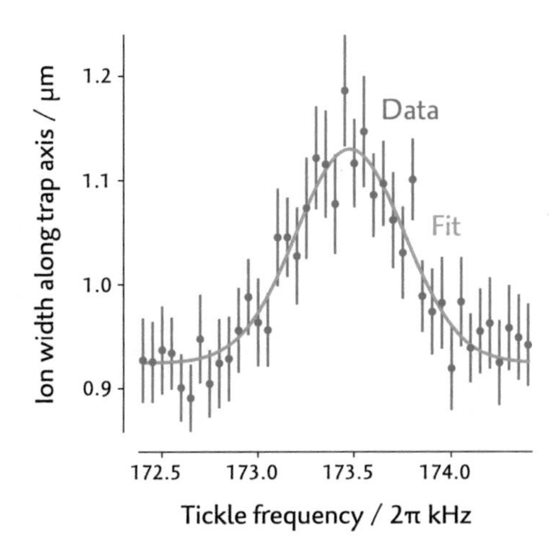

Fig. 5.3 A driving field causes the ion crystal to heat if the field is resonant with a motional mode. The ions then become diffuse in the image. The data displayed is generated by fitting image data by a two-dimensional Gaussian function; error bars indicate fit uncertainties (68% confidence interval). This data is in turn fit by the Gaussian function shown in green; motional mode frequencies are then determined to 1 part in 7000

The motional frequencies can be determined from the frequencies of motional sidebands in the $5S_{1/2} \leftrightarrow 4D_{5/2}$ spectrum, however this requires ac-Stark shifts to be carefully accounted for. Instead we use the 'tickle' method [17]: a radiofrequency (RF) voltage is applied to the endcap electrodes, if it is resonant with a motional mode in the axial direction the axial motion is driven and the width of the ions in the EMCCD camera image is increased. The widths of the ions are recorded as the frequency of the 'tickle' voltage is scanned and in this fashion the motional mode frequencies are accurately measured. Typical results are shown in Fig. 5.3.

References

1. Feldker T (2017) Rydberg excitation of trapped ions. Ph.D thesis, Johannes Gutenberg-Universität Mainz
2. Müller M, Liang L, Lesanovsky I, Zoller P (2008) Trapped Rydberg ions: from spin chains to fast quantum gates. New J Phys 10:093009
3. Leibrandt DR et al (2007) Laser ablation loading of a surfaceelectrode ion trap. Phys Rev A 76:055403
4. Łabaziewicz J (2008) High fidelity quantum gates with ions in cryogenic microfabricated ion traps. Ph.D thesis, Massachusetts Institute of Technology
5. Kielpinski D, Monroe C, Wineland DJ (2002) Architecture for a large-scale ion-trap quantum computer. Nature 417:709
6. Ofek N et al (2016) Extending the lifetime of a quantum bit with error correction in superconducting circuits. Nature 536:441
7. Browaeys A, Barredo D, Lahaye T (2016) Experimental investigations of dipole-dipole interactions between a few Rydberg atoms. J Phys B 49:152001
8. Saffman M (2016) Quantum computing with atomic qubits and Rydberg interactions: progresss s and challenges. J Phys B 49:202001
9. Saffman M, Walker TG (2005) Analysis of a quantum logic device based on dipole-dipole interactions of optically trapped Rydberg atoms. Phys Rev A 72:022347
10. Potvliege RM, Adams CS (2006) Photo-ionization in far-offresonance optical lattices. New J Phys 8:163

11. Johnson TA et al (2008) Rabi oscillations between ground and Rydberg states with dipole-dipole atomic interactions. Phys Rev Lett 100:113003
12. Saffman M, Walker TG, Mølmer K (2010) Quantum information with Rydberg atoms. Rev Mod Phys 82:2313–2363
13. Beterov II, Tretyakov DB, Ryabtsev II, Ekers A, Bezuglov NN (2007) Ionization of sodium and rubidium nS, nP, and nD Rydberg atoms by blackbody radiation. Phys Rev A 75:052720
14. James D (1998) Quantum dynamics of cold trapped ions with application to quantum computation. Appl Phys B 66:181–190
15. Home JP, Hanneke D, Jost JD, Leibfried D, Wineland DJ (2011) Normal modes of trapped ions in the presence of anharmonic trap potentials. New J Phys 13:073026
16. Audi G, Bersillon O, Blachot J, Wapstra A (2003) The Nubase evaluation of nuclear and decay properties. Nucl Phys A 729:3–128
17. Nägerl HC, Leibfried D, Schmidt-Kaler F, Eschner J, Blatt R (1998) Coherent excitation of normal modes in a string of Ca^+ ions. Opt Express 3:89–96

Chapter 6
Trap Effects

Rydberg states are extremely sensitive to electric fields. In the seminal theoretical investigation by Müller et al. [1] effects of strong electric trapping fields on highly-sensitive Rydberg ions were predicted. Experimental observation of these effects is presented here. In this chapter we also investigate a Rydberg ion in trapping regimes beyond the consideration of Müller et al.; both theoretically and experimentally.

Two classes of trap effects emerge, one related to the Rydberg ion electric polarisability, the other related to the Rydberg ion electric quadrupole moment.

The Rydberg state polarisability scales as n^7, this makes Rydberg atoms sensitive electric field probes [2]. Effects of the trapping electric fields on highly-polarisable Rydberg ions are explored in Sect. 6.1.

Rydberg states with $J > \frac{1}{2}$ have large electric quadrupole moments, which scale as n^4. Effects of the trapping electric quadrupole fields on states with large electric quadrupole moments (in particular a Rydberg $D_{3/2}$ state) are explored in Sect. 6.2. This second class of effects is negligible for $J = \frac{1}{2}$ states, namely $nS_{1/2}$ and $nP_{1/2}$, with $n < 50$ [1].

6.1 Effects on Highly-Polarisable Rydberg Ions

Owing to their large polarisabilities, Rydberg ions experience a different trapping pseudopotential than ions in low-lying states. This leads to different effects, which depend on whether or not the nulls of the oscillating and static electric quadrupole fields overlap.

Trap effects were theoretically studied in [1] for a trap in which the electric field nulls overlap. This corresponds to the case with no excess micromotion. We experimentally investigated these effects [3, 4], the results are presented in Sect. 6.1.1.

We have also experimentally investigated the effects which appear when the field nulls do not overlap [4], and explain the results by extending the theory. When the

G. Higgins, *A Single Trapped Rydberg Ion*, Springer Theses,
https://doi.org/10.1007/978-3-030-33770-4_6

field nulls do not overlap there is excess micromotion in the system. This work is described in Sect. 6.1.2.

6.1.1 With Overlapping Quadrupole Field Nulls

Due to the large Rydberg state electric polarisability, the radial trapping frequencies of a Rydberg ion are different to the radial trapping frequencies of an ion in a low-lying state. This means the energy required for Rydberg excitation depends on the number of phonons in radial motional modes. Before the experimental results are presented, this effect is described using a semi-classical approach. A full quantum mechanical description is found in [1, 3, 5].

Description Using Semi-classical Theory

Near the centre of a linear Paul trap the electric trapping potential Φ is composed of two electric quadrupole potentials, one which oscillates and one which is static:

$$\Phi = \alpha \cos \Omega_{\mathrm{rf}} t (x^2 - y^2) - \beta[(1+\epsilon)x^2 + (1-\epsilon)y^2 - 2z^2], \tag{6.1}$$

where α is the time-dependent field gradient which oscillates with radiofrequency Ω_{rf}, β is the static field gradient, and the parameter ϵ breaks the axial symmetry of the trap and lifts the radial mode degeneracy.

In the axial direction the trapping potential is harmonic with frequency ω_z given by

$$\omega_z^2 = \frac{4e\beta}{M}, \tag{6.2}$$

where e is the elementary charge (i.e. the ion charge) and M is the ion mass.

The radial motion of an ion near the centre of the trapping potential is described in terms of the Mathieu equation. When $\beta \ll \alpha \ll \frac{M\Omega_{\mathrm{rf}}^2}{4e}$ the first-order solution describes ion motion in each radial direction composed of two parts: harmonic motion with frequency ω_x, ω_y called secular motion and driven motion with frequency Ω_{rf} called micromotion, where

$$\begin{aligned} \omega_x^2 &= 2\left(\frac{e\alpha}{M\Omega_{\mathrm{rf}}}\right)^2 - \frac{2e\beta\,(1+\epsilon)}{M}, \\ \omega_y^2 &= 2\left(\frac{e\alpha}{M\Omega_{\mathrm{rf}}}\right)^2 - \frac{2e\beta\,(1-\epsilon)}{M}. \end{aligned} \tag{6.3}$$

The secular approximation involves neglecting the fast oscillating micromotion and interpreting the secular motion as generated by a time-independent harmonic pseudopotential

$$U_{\text{trap}} = \tfrac{1}{2}M(\omega_x^2 x^2 + \omega_y^2 y^2 + \omega_z^2 z^2). \tag{6.4}$$

The equilibrium position of a trapped ion is $\vec{r}_0 = \begin{pmatrix}0 & 0 & 0\end{pmatrix}$; at this position both electric quadrupole potentials have saddle points and thus both electric quadrupole fields have nulls.

Rydberg ions have giant polarisabilities ρ which scale with the principal quantum number as n^7. In an electric field of strength $\mathcal{E}$ a Rydberg ion experiences an energy shift due to the quadratic Stark effect

$$\Delta E = -\tfrac{1}{2}\rho\mathcal{E}^2. \tag{6.5}$$

From Eq. (6.1) the trapping electric field is given by

$$\vec{\mathcal{E}} = -\vec{\nabla}\Phi = -2\alpha\cos\Omega_{\text{rf}}t\begin{pmatrix}x\\-y\\0\end{pmatrix} + 2\beta\begin{pmatrix}(1+\epsilon)x\\(1-\epsilon)y\\-2z\end{pmatrix}. \tag{6.6}$$

Rydberg excitation is carried out over a much longer timescale than the trap period $[\left(\frac{\Omega_{\text{eff}}}{2\pi}\right)^{-1} \gg \left(\frac{\Omega_{\text{rf}}}{2\pi}\right)^{-1} = 55\,\text{ns}$ in this system] and thus we time-average the squared electric field strength, and use $\alpha^2 \gg 2\beta^2$ to find

$$\langle\mathcal{E}^2\rangle \approx 2\alpha^2\left(x^2 + y^2\right). \tag{6.7}$$

And thus in addition to the trap pseudopotential [Eq. (6.4)] the highly-polarisable Rydberg ion experiences the additional harmonic potential

$$U_{\text{add}} = -\frac{1}{2}\rho\langle\mathcal{E}^2\rangle \approx -\rho\alpha^2(x^2 + y^2), \tag{6.8}$$

and thus the radial trapping frequencies become

$$\begin{aligned}\omega_x \to \omega_x' &\approx \sqrt{\omega_x^2 - \frac{2\rho\alpha^2}{M}},\\ \omega_y \to \omega_y' &\approx \sqrt{\omega_y^2 - \frac{2\rho\alpha^2}{M}},\end{aligned} \tag{6.9}$$

while the axial trapping frequency is still described by Eq. (6.2), because the z-dependence in Eq. (6.7) is negligible.

The energy for Rydberg excitation is thus shifted depending on radial phonon numbers

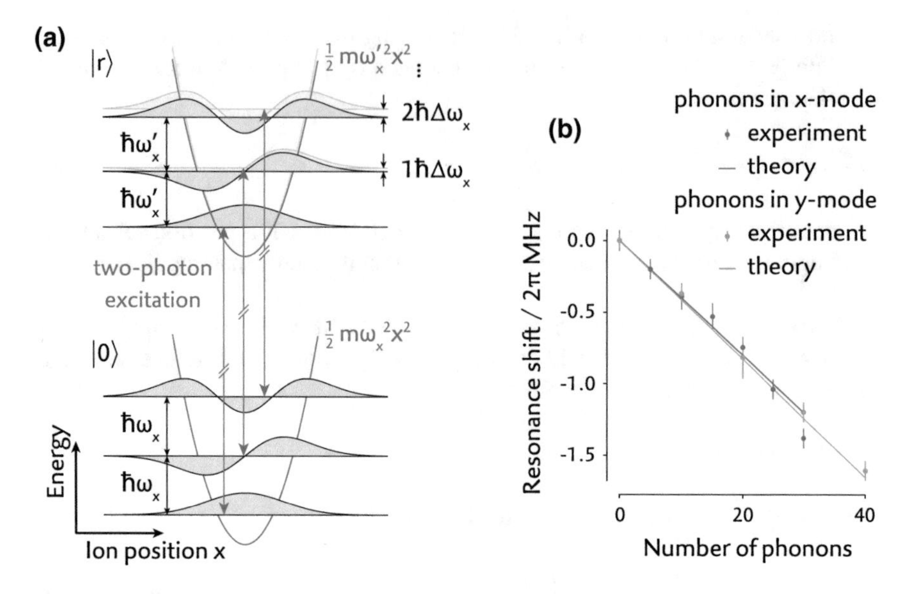

Fig. 6.1 Rydberg ions experience an altered trapping potential, and thus the frequency of the Rydberg excitation transition depends linearly on the number of phonons in motional modes. **a** The trapping pseudopotentials are in blue, Rydberg excitation transitions are in purple. **b** Measurement results and theory show excellent agreement; the theory returns $\Delta\omega_x = -2\pi \times 40\,\text{kHz}$ and $\Delta\omega_y = -2\pi \times 41\,\text{kHz}$ for $46S_{1/2}$ and the trapping parameters used. Error bars are dominated by laser frequency drifts (68% confidence interval)

$$\Delta E = n_x \hbar \Delta\omega_x + n_y \hbar \Delta\omega_y, \tag{6.10}$$

where

$$\begin{aligned} \Delta\omega_x &= \omega'_x - \omega_x, \\ \Delta\omega_y &= \omega'_y - \omega_y, \end{aligned} \tag{6.11}$$

as sketched in Fig. 6.1a.

The altered trapping pseudopotential also causes phonon-number-changing transitions to be driven more strongly during Rydberg excitation. This effect is more noticeable when the nulls of the quadrupole trapping fields do not overlap, and it is discussed in Sect. 6.1.2.

Experimental Results

The dependence of the energy required for Rydberg excitation on the number of radial phonons [Eq. (6.10)] is observed directly by preparing an ion with a known number of radial phonons (n_x and n_y) and measuring the Rydberg resonance frequency. The

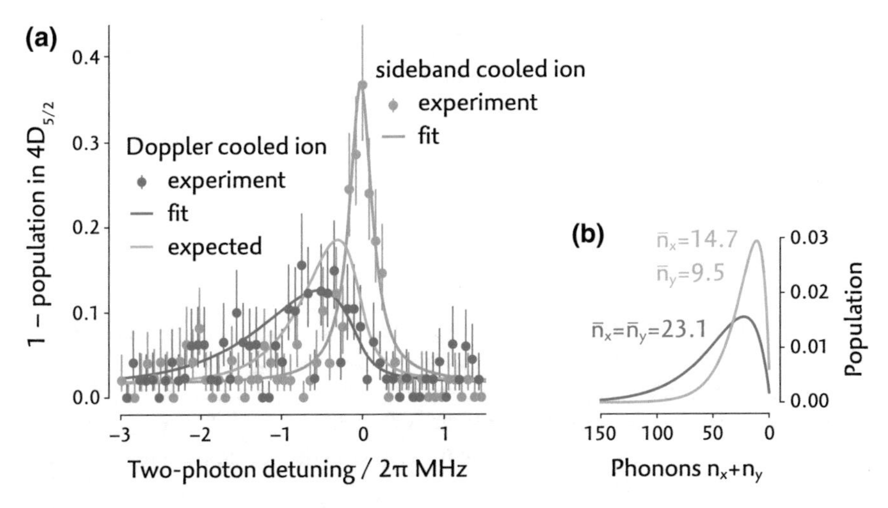

Fig. 6.2 **a** Because the Rydberg excitation transition frequency depends on the number of phonons in radial modes, the Rydberg resonance linewidth is broader for a Doppler-cooled ion (with broad phonon distributions) than a sideband-cooled ion (with narrow phonon distributions). The fit of the Doppler cooled ion resonance lineshape reflects thermal phonon distributions with $\bar{n}_x = \bar{n}_y = 23.1 \pm 2.1$. The expected lineshape of the Doppler cooled ion resonance does not match the data as well, it uses independent estimates of $\bar{n}_x = 14.7 \pm 0.9$, $\bar{n}_y = 9.5 \pm 0.8$ and no fit parameters. The $42S_{1/2}$ polarisability and the trap parameters used result in $\Delta\omega_x \approx \Delta\omega_y = -2\pi \times 20\,\text{kHz}$. Error bars indicate quantum projection noise (68% confidence interval). Distributions of the total number of radial phonons are shown in **b**

results in Fig. 6.1b are consistent with Eq. (6.10). Preparation of phonon-number Fock states is described in Sect. 3.1.3.

Because of this effect the Rydberg resonance linewidth depends on the widths of the radial phonon distributions. As shown in Fig. 6.2a, a narrow Rydberg resonance is obtained when a sideband cooled ion (with narrow phonon distributions) is excited, while a broader Rydberg resonance is observed when a Doppler cooled ion (with broader phonon distributions) is excited. The resonance lineshape obtained with a Doppler cooled ion reflects the population distribution over the two radial modes, shown in Fig. 6.2b. Radial sideband cooling is important for coherent excitation of Rydberg states, because the efficiency of coherent excitation depends on the Rydberg state linewidth (see Sect. 7.3.2).

6.1.2 *With Non-overlapping Quadrupole Field Nulls*

Stray electric fields can cause the electric quadrupole field nulls to be separated [6], and excess micromotion results from this (see Sect. 3.2.8). When the electric quadrupole field nulls do not overlap additional effects emerge on a trapped Rydberg ion: phonon-number-changing transitions are driven more strongly during Rydberg excitation, and the energy required for Rydberg excitation depends on the distance between the field nulls. These effects stem from the altered trapping pseudopotential of the Rydberg ion.

Before the experimental observation of these effects is shown, the theory underlying these effects is presented.

Background Theory

In this regime the null of the oscillating electric quadrupole field ($\vec{r}_{rf}$) and the null of the static electric quadrupole field ($\vec{r}_{dc}$) do not overlap ($\vec{r}_{rf} \neq \vec{r}_{dc}$). Then the equilibrium position of an ion in a low-lying state ($\vec{r}_g$) does not correspond to either null ($\vec{r}_g \neq \vec{r}_{rf}, \vec{r}_{dc}$). Due to the large Rydberg state electric polarisability the equilibrium position of the Rydberg ion ($\vec{r}_R$) does not overlap with the equilibrium position of an ion in a low-lying state ($\vec{r}_R \neq \vec{r}_g$).

When the oscillating component of the trapping potential is centred along the z-axis $\vec{r}_{rf} = \begin{pmatrix} 0 & 0 & z \end{pmatrix}$ and the static trapping potential is centred on $\vec{r}_{dc} = \begin{pmatrix} x_{dc} & y_{dc} & 0 \end{pmatrix}$ the total electric trapping potential is modified from Eq. (6.1) and becomes

$$\Phi = \alpha \cos \Omega_{rf} t (x^2 - y^2) - \beta[(1+\epsilon)(x - x_{dc})^2 + (1-\epsilon)(y - y_{dc})^2 - 2z^2]. \tag{6.12}$$

The oscillating and static electric quadrupole fields (given by $\vec{\mathcal{E}} = -\vec{\nabla}\Phi$) then have nulls which do not overlap. The separation between $\vec{r}_{rf}$ and $\vec{r}_{dc}$ may be introduced by a static electric field

$$\vec{\mathcal{E}}_{offset} = -2\beta \begin{pmatrix} (1+\epsilon)x_{dc} \\ (1-\epsilon)y_{dc} \\ 0 \end{pmatrix}. \tag{6.13}$$

For an ion in a low-lying state the trapping pseudopotential is then centred on $\vec{r}_g = \begin{pmatrix} x_g & y_g & 0 \end{pmatrix}$ [6]

$$U_{trap} = \frac{1}{2} M(\omega_x^2 (x - x_g)^2 + \omega_y^2 (y - y_g)^2 + \omega_z^2 z^2), \tag{6.14}$$

where[1]

$$\begin{aligned} x_g &= -\frac{2e\beta(1+\epsilon)\, x_{dc}}{M\omega_x^2}, \\ y_g &= -\frac{2e\beta(1-\epsilon)\, y_{dc}}{M\omega_y^2}. \end{aligned} \tag{6.15}$$

[1]Generally $\omega_x \neq \omega_y$ and $\epsilon \neq 0$ and thus $\vec{r}_{rf}$, $\vec{r}_{dc}$ and $\vec{r}_g$ are not collinear.

Owing to its high polarisability, a Rydberg ion experiences an additional harmonic potential which is centred on $\vec{r}_{\rm rf}$ [since the approximation in Eq. (6.7) still holds]

$$\begin{aligned} U'_{\rm trap} &= U_{\rm trap} - \frac{1}{2}\rho\langle\mathcal{E}^2\rangle \\ &\approx U_{\rm trap} - \rho\alpha^2(x^2+y^2). \end{aligned} \tag{6.16}$$

After some simple algebra, we find the trapping pseudopotential of a Rydberg ion is centred on $\vec{r}_{\rm R} = \begin{pmatrix} x_{\rm R} & y_{\rm R} & 0 \end{pmatrix}$, and it is shifted in energy (relative to the pseudopotential of the ion in a low-lying electronic state) by ΔU

$$U'_{\rm trap} = \tfrac{1}{2}M(\omega_x'^2(x-x_{\rm R})^2 + \omega_y'^2(y-y_{\rm R})^2 + \omega_z^2 z^2) + \Delta U, \tag{6.17}$$

where

$$\begin{aligned} x_{\rm R} &= x_{\rm g}\left(1 - \frac{2\rho\alpha^2}{M\omega_x^2}\right)^{-1}, \\ y_{\rm R} &= y_{\rm g}\left(1 - \frac{2\rho\alpha^2}{M\omega_y^2}\right)^{-1}, \end{aligned} \tag{6.18}$$

$$\begin{aligned} \Delta U &= \tfrac{1}{2}M\left(\omega_x^2 x_{\rm g}^2 + \omega_y^2 y_{\rm g}^2 - \omega_x'^2 x_{\rm R}^2 - \omega_y'^2 y_{\rm R}^2\right) \\ &\approx -\rho\alpha^2(x_{\rm g}^2 + y_{\rm g}^2) \\ &\approx -\rho\alpha^2(x_{\rm R}^2 + y_{\rm R}^2) = -\tfrac{1}{2}\rho\langle\mathcal{E}(\vec{r}_{\rm R})^2\rangle, \end{aligned} \tag{6.19}$$

and ω_x' and ω_y' are unchanged from Eq. (6.9). The approximations in Eq. (6.19) are valid when $\frac{2\rho\alpha^2}{M\omega_x^2} \ll 1$ and $\frac{2\rho\alpha^2}{M\omega_y^2} \ll 1$. From Eq. (6.19) we understand that the energy shift ΔU results from the electric field at $\vec{r}_{\rm R}$ acting on the highly-polarisable Rydberg ion.

Due to differences between the equilibrium positions $\vec{r}_{\rm g}$, and $\vec{r}_{\rm R}$ phonon-number-changing transitions may be driven strongly. Rydberg resonance frequency shifts result from ΔU. These phenomena (which stem from the Rydberg ion polarisability) are presented in the following subsections.

Phonon-Number-Changing Transitions

The motional components of the total ion wavefunction are described by the eigenstates of the quantum harmonic oscillator. These eigenstates are symmetric about the pseudopotential centre and they depend on the phonon number and the pseudopotential frequency. Due to the difference in trapping pseudopotentials, the radial motional wavefunctions of ions in low-lying electronic states $|n_x\rangle$, $|n_y\rangle$ and ions in Rydberg

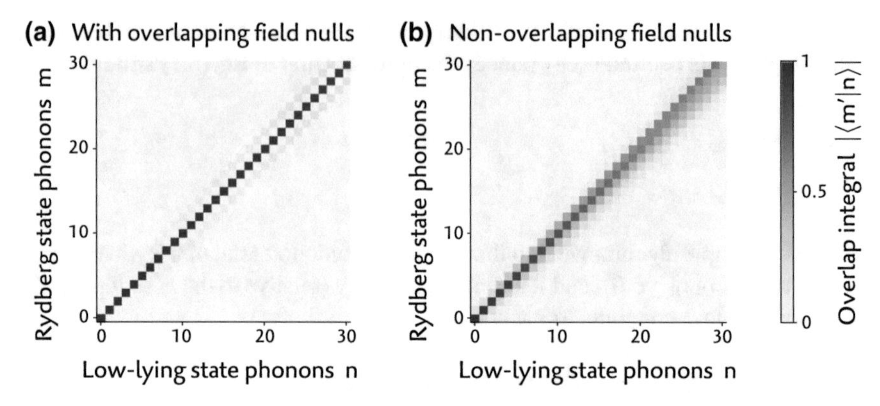

Fig. 6.3 Phonon-number-changing transitions are driven more strongly during Rydberg excitation when the number of phonons is higher and when the field nulls do not overlap. The strength is given by the overlap of motional mode states (Franck–Condon factors). Here a single radial mode is considered, and the trapping frequencies obey $\omega' = 0.975 \times \omega$. In **a** only transitions which change the phonon-number by an even number are driven, due to symmetry. In **b** the pseudopotential minima of a Rydberg state and a low-lying state are separated by $0.2\sqrt{\frac{\hbar}{M\omega}}$. During typical operation of the experiment we expect this separation to be $\sim 0.04\sqrt{\frac{\hbar}{M\omega}}$ as a result of imperfect minimisation of micromotion

states $|m'_x\rangle$, $|m'_y\rangle$ do not overlap perfectly $\langle m'_x|n_x\rangle \neq \delta_{m_x n_x}$, $\langle m'_y|n_y\rangle \neq \delta_{m_y n_y}$. As a result phonon-number-changing transitions may be driven during Rydberg excitation, with strengths described by the overlap of the motional wavefunctions. The overlap integrals are called Franck–Condon factors.[2]

Phonon-number-changing transitions are expected both when the trapping field nulls overlap and when the nulls do not overlap. When the nulls do not overlap the Franck–Condon factors may be significantly higher and phonon-number-changing transitions may be driven more strongly. This is shown in Fig. 6.3.

In the experiment we introduce an additional static electric field [Eq. (6.13)] so that the saddle points of the electric quadrupole potentials in Eq. (6.12) do not overlap ($\vec{r}_{\rm rf} \neq \vec{r}_{\rm dc}$). This means the pseudopotential centres do not overlap ($\vec{r}_{\rm R} \neq \vec{r}_{\rm g}$) and strong phonon-number-changing transitions are observed in the Rydberg excitation spectrum, shown in Fig. 6.4.

Using ΔU to Minimise Micromotion

Ion micromotion is introduced in Sect. 3.2.8. The amplitude of micromotion depends on the separation of the ion equilibrium position and the RF field null $|\vec{r}_{\rm g} - \vec{r}_{\rm rf}|$ [6].

[2]Since the Rydberg-excitation lasers counter-propagate (Sect. 4.4), the contribution to phonon-number-changing transitions described by the Lamb–Dicke parameter is small compared with the Franck–Condon factors considered in this section.

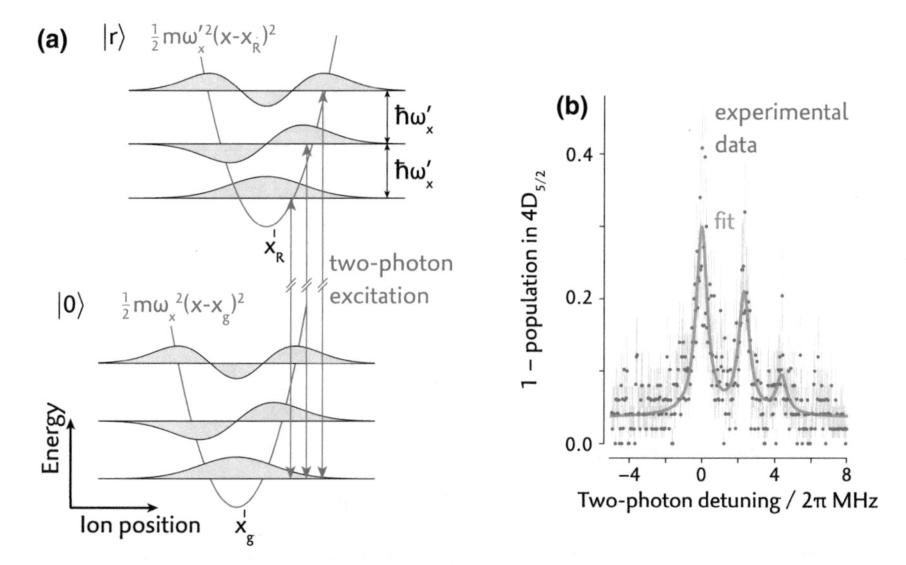

Fig. 6.4 Phonon-number-changing transitions due to non-overlapping pseudopotential minima: **a** Because the minima of the trapping pseudopotentials for low-lying states and for Rydberg states do not overlap, the motional modes overlap poorly and phonon-number-changing transitions may be driven strongly during Rydberg excitation. **b** When the field nulls do not overlap the Rydberg resonance structure of a sideband cooled ion shows strong blue sidebands displaced from the carrier transition by $+\omega'$ and $+2\omega'$, which correspond to addition of 1 and 2 radial photons. The radial trapping frequencies used are $\omega_{x,y} \approx \omega'_{x,y} \approx 2\pi \times 2.1$ MHz. Error bars indicate quantum projection noise (68% confidence interval)

Micromotion is minimised by introducing static electric fields such that the distances between $\vec{r}_{\mathrm{rf}}$ and $\vec{r}_{\mathrm{dc}}$ (and thus $\vec{r}_{\mathrm{g}}$ and $\vec{r}_{\mathrm{R}}$) are minimised. The linear Paul trap has two electrodes used to introduce these fields [see Fig. 3.3b]. Appropriate compensation fields are found using a range of established techniques, described in Sect. 3.2.8. Here we introduce a novel method for micromotion minimisation: The separation $|\vec{r}_{\mathrm{g}} - \vec{r}_{\mathrm{rf}}|$ causes a shift in the energy required for Rydberg excitation ΔU [see Eq. (6.17)]; by minimising this energy shift, micromotion is minimised.

This method is demonstrated in Fig. 6.5a. Here the Rydberg resonance frequency varies quadratically with the voltage applied to a micromotion-compensation electrode. This is consistent with Eq. (6.19), since the distance $|\vec{r}_{\mathrm{g}} - \vec{r}_{\mathrm{rf}}|$ varies linearly with the voltage [according to Eqs. (6.13), (6.15)].[3]

This method for minimising micromotion is limited by the sensitivity of Rydberg states to electric fields and by frequency drifts of the Rydberg-excitation lasers. Laser frequency drifts ($\sim 2\pi \times 300\,\mathrm{Hz\,s^{-1}}$) limit the accuracy with which resonance frequencies are determined and thus the accuracy with which ΔU is determined to $\sim\hbar\, 2\pi \times 100\,\mathrm{kHz}$. Using Rydberg state $46S_{1/2}$ (with polarisability

[3]The linear relation holds for one radial direction provided that the ion is close to the trap centre in the orthogonal radial direction.

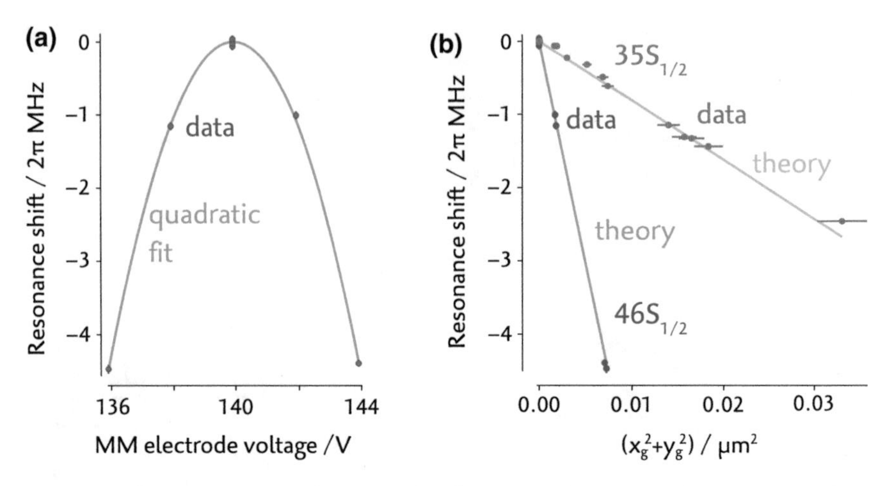

Fig. 6.5 The Rydberg resonance frequency depends on the distance between the electric quadrupole field nulls. **a** The Rydberg resonance frequency varies quadratically with the voltages applied to micromotion compensation electrodes. The maximum resonance frequency corresponds to the case with overlapping field nulls (when micromotion is minimised). Here the horizontal micromotion compensation electrode voltage is varied. Error bars are barely discernible on this graph, they result from laser frequency drifts (68% confidence interval). **b** The Rydberg resonance frequency shift is proportional to the squared distance between the ground state ion equilibrium position and the RF field null, with gradient $-\rho\alpha^2/\hbar$ [according to Eq. (6.19)]. The gradient thus depends on the Rydberg state polarisability. The horizontal error bars result from uncertainty in the conversion between voltage and distance (68% confidence interval)

$\rho_{46S} = 6 \times 10^{-31}\,\mathrm{C\,m^2\,V^{-1}}$) the residual oscillating electric field strength at the equilibrium position of the ion $\vec{r}_g$ is reduced to $\mathcal{E}_{rf} \sim 20\,\mathrm{V\,m^{-1}}$. With the best micromotion minimisation techniques the residual field strength can be reduced to $\mathcal{E}_{rf} \sim 0.3\,\mathrm{V\,m^{-1}}$ [7].

This method may allow significantly lower residual field strengths to be reached if the frequency stability of the UV lasers was improved and if Rydberg states with higher polarisabilities were used. With laser frequency drifts $<2\pi \times 1\,\mathrm{Hz\,s^{-1}}$ (this is experimentally realisable) ΔU may be determined with precision $\sim\hbar\, 2\pi \times 1\,\mathrm{kHz}$ (provided the Rydberg state resonance is narrow enough). $nP_{1/2}$ states have polarisabilities $\sim$5 times higher than $nS_{1/2}$ states and have been excited in our lab (see the Outlook in Chap. 8). State $59P_{1/2}$ has linewidth $\Gamma \approx 2\pi \times 2\,\mathrm{kHz}$ at 300 K and polarisability $\rho \approx -10^{10}\,\mathrm{C\,m^2\,V^{-1}}$. By using $59P_{1/2}$ as a probe and with more stable frequency references residual field strengths may be reduced to $\mathcal{E}_{rf} \sim 0.3\,\mathrm{V\,m^{-1}}$.

In Fig. 6.5b we confirm Eq. (6.19) is followed quantitatively. The voltage axis in Fig. 6.5a is converted to a distance in Fig. 6.5b using the EMCCD camera image of a trapped ion. This method relies on knowledge of the image magnification, which is measured using two ions as described in Sect. 5.2.1.

6.2 Effects on Rydberg Ions with Large Quadrupole Moments

Rich resonance structures are observed when Rydberg D-states are excited. The structures include Floquet sidebands, which result from coupling between levels driven by the time-dependent electric quadrupole trapping field. The strength of this coupling depends on the electric quadrupole moments of the states in question; effects of the coupling are best observed with $J > \frac{1}{2}$ states (such as $nD_{3/2}$) which have sizeable quadrupole moments.

The coupling is negligible for $J = \frac{1}{2}$ states ($nS_{1/2}$ and $nP_{1/2}$) with $n < 50$ [1], since $J = \frac{1}{2}$ states have negligible quadrupole moments and the energy separation between $J = \frac{1}{2}$ and $J > \frac{1}{2}$ states makes coupling between them weak. The experiments proposed in Sect. 1.3.1 are concerned with Rydberg $S_{1/2}$- and $P_{1/2}$-states because they are simpler systems.

The theory behind the trap-induced coupling is presented in Sect. 6.2.1. The observed effects of this coupling are understood in terms of Floquet theory; this theory is introduced in Sect. 6.2.2. Experimental results are shown in Sect. 6.2.3.

6.2.1 *Theory Behind the Trap-Induced Coupling*

The Rydberg electron interacts with the trapping electric quadrupole fields according to

$$H_{\mathrm{eQ}} = -e\Phi \tag{6.20}$$

$$= e\beta\left(x^2 + y^2 - 2z^2\right) - e\alpha\cos\Omega_{\mathrm{rf}}t\left(x^2 - y^2\right) \tag{6.21}$$

$$= -4\sqrt{\frac{\pi}{5}}er^2\beta Y_2^0 - 2\sqrt{\frac{2\pi}{15}}er^2\alpha\cos\Omega_{\mathrm{rf}}t\left(Y_2^2 + Y_2^{-2}\right) \tag{6.22}$$

$$= \sum_q \underbrace{A_q}_{\substack{\mathcal{E}\text{-field}\\ \text{gradient}}} \times \underbrace{er^2 Y_2^q}_{\substack{\text{spherical electric}\\ \text{quadrupole moment}}}, \tag{6.23}$$

where for clarity we define

$$A_0 = -4\sqrt{\frac{\pi}{5}}\beta, \quad A_{\pm 2} = -2\sqrt{\frac{2\pi}{15}}\alpha\cos\Omega_{\mathrm{rf}}t. \tag{6.24}$$

We use the trapping potential from Eq. (6.1) and neglect the trap anisotropy ϵ. By writing H_{eQ} in terms of spherical harmonics we see H_{eQ} is composed of spherical electric quadrupole moments and electric field gradients.

For Rydberg D-states the strength of H_{eQ} (which scales as n^4) is much weaker than the fine structure splitting ($\sim n^{-3}$) until $n \approx 45$. We experimentally investigate $24D_{3/2}$ and $27D_{3/2}$, and thus to describe the results we need only consider the manifold of states described by quantum numbers $\{n, L, S, J\}$ which have different m_J.

In our experiment the trap axis coincides with the quantisation axis (defined by the magnetic field, see Sect. 3.2.6) and the matrix elements of the components of H_{eQ} are

$$
\begin{aligned}
&\langle nLSJm'_J|A_q e r^2 Y_2^q|nLSJm_J\rangle \\
&\quad = A_q e \langle nLJ|r^2|nLJ\rangle\langle LSJm'_J|Y_2^q|LSJm_J\rangle && (6.25)\\
&\quad = (-1)^{J-m'_J} A_q e\langle r^2\rangle (LSJ||Y_2||LSJ) \begin{pmatrix} J & 2 & J \\ -m'_J & q & m_J \end{pmatrix} && (6.26)\\
&\quad = (-1)^{J-m'_J+1}\frac{1}{2}\sqrt{\frac{5}{\pi}} A_q \Theta(nLSJ) \\
&\qquad \times \begin{pmatrix} J & 2 & J \\ -J & 0 & J \end{pmatrix}^{-1} \begin{pmatrix} J & 2 & J \\ -m'_J & q & m_J \end{pmatrix}. && (6.27)
\end{aligned}
$$

In Eq. (6.25) we make use of the separability of the atomic wavefunction and in Eqs. (6.26), (6.27) the Wigner–Eckard theorem is used. The electric quadrupole moment of $|nLSJ\rangle$ is defined by the diagonal matrix element of the state with $m_J = J$ [8]

$$
\begin{aligned}
\Theta(nLSJ) &= -\frac{e}{2}\langle nLSJJ|3z^2 - r^2|nLSJJ\rangle && (6.28)\\
&= -2\sqrt{\frac{\pi}{5}} e\langle r^2\rangle\langle SLJJ|Y_2^0|SLJJ\rangle && (6.29)\\
&= -2\sqrt{\frac{\pi}{5}} e\langle r^2\rangle (SLJJ||Y_2||SLJJ) \begin{pmatrix} J & 2 & J \\ -J & 0 & J \end{pmatrix}. && (6.30)
\end{aligned}
$$

Whenever the trap axis and the quantisation axis do not coincide, the matrix elements of H_{eQ} are found by rotating between the bases set by the trap axis and the quantisation basis. In the experiment the axes coincide and so we do not need to rotate between bases here.

Selection rules become apparent through the properties of Wigner-3j symbols. The electric quadrupole moment $\Theta(nLSJ)$ and the matrix elements of Eq. (6.27) are zero when the triangle inequality $|J - 2| \leq J$ is not satisfied, i.e. for $J = \frac{1}{2}$ states. This is why we do not observe quadrupole effects in the excitation spectra of Rydberg $S_{1/2}$-states. Non-zero matrix elements must satisfy $m'_J = m_J + q$ and thus the $q = 0$ term has only on-diagonal elements which cause levels to be shifted in energy, while the $q = \pm 2$ terms have only off-diagonal elements which couple states which differ in m_J by 2. The coupling oscillates with frequency Ω_{rf}.

Hamiltonian H_{eQ} is not only relevant in trapped Rydberg ion experiments; trapped ion atomic clocks must account for quadrupole shifts in transitions frequencies which result from H_{eQ} [8]. Perturbation theory is used to describe these shifts of $\sim 2\pi \times 10\,\text{Hz}$. The effects of H_{eQ} are more dramatic in Rydberg states because $\Theta(nLSJ)$ scales with n^4, for example $\Theta(27D_{3/2}) = 5.3 \times 10^4\, e\, a_0{}^2$ (calculated using theoretical wavefunctions, see Sect. 2.2), while $\Theta(4D_{5/2}) = 3.0\, e\, a_0{}^2$ [9] and $\Theta(4D_{3/2}) = 2.0\, e\, a_0{}^2$ [10].

The $q = 0$ term of H_{eQ} causes the Zeeman sublevels of a $nD_{3/2}$ state to be shifted according to

$$E_s = (-1)^{|m_J|+1/2} 2\beta\Theta(nD_{3/2}). \tag{6.31}$$

This formula holds for a single trapped ion in the absence of micromotion. When multiple ions are trapped the Coulombic fields contribute to the static electric quadrupole fields [11].

The $q = \pm 2$ components of H_{eQ} can be written as

$$\begin{aligned} H_{\text{rf}} = \hbar C \cos \Omega_{\text{rf}} t \sum_{m_J=1/2}^{3/2} & |nD_{3/2}(m_J - 2)\rangle\langle nD_{3/2} m_J| \\ & + |nD_{3/2}(m_J + 2)\rangle\langle nD_{3/2} m_J|, \end{aligned} \tag{6.32}$$

where

$$C = \frac{2\alpha\Theta(nD_{3/2})}{\sqrt{3}\hbar}. \tag{6.33}$$

The two effects are represented in Fig. 6.6.

When we excite $24D_{3/2}$ and $27D_{3/2}$ Rydberg states in the laboratory we observe rich resonance structures containing Floquet sidebands. The sidebands appear because the coupling strength C is comparable with the trap drive Ω_{rf} and thus effects of H_{rf} extend beyond what may be described within the rotating wave approximation. Floquet theory is introduced next, before the experimental results are shown in Sect. 6.2.3.

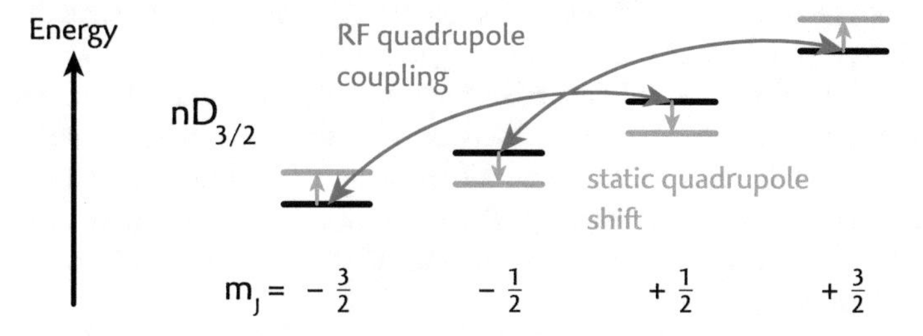

Fig. 6.6 The static electric quadrupole field shifts the energies of Rydberg $D_{3/2}$-state sublevels. When the trap axis and magnetic field are collinear the RF oscillating field couples next-neighbouring sublevels. In the absence of trap effects sublevels are split because of the Zeeman effect

6.2.2 Floquet Theory

The energy spectrum of a periodically-driven system, with drive frequency Ω, shows sidebands at integer multiples of $\hbar\Omega$. These sidebands may be understood in terms of Floquet theory, which is outlined here following the approach in [12].

The Hamiltonian of a periodically-driven system satisfies

$$H(\vec{r},t) = H(\vec{r},t+T), \tag{6.34}$$

where $T = \frac{2\pi}{\Omega}$. The periodicity of H suggests looking for solutions of the time-dependent Schrödinger equation $\psi(\vec{r},t)$ with the same periodicity (to within a phase factor)

$$\psi(\vec{r},t) = e^{-i\epsilon t/\hbar}\phi_\epsilon(\vec{r},t), \tag{6.35}$$

where

$$\phi_\epsilon(\vec{r},t) = \phi_\epsilon(\vec{r},t+T). \tag{6.36}$$

By substituting Eq. (6.35) into the time-dependent Schrödinger equation $H\psi = i\hbar\frac{\partial\psi}{\partial t}$ we see

$$\left(H - i\hbar\frac{\partial}{\partial t}\right)\phi_\epsilon = \epsilon\phi_\epsilon \tag{6.37}$$

and thus ϕ_ϵ is an eigenstate of

$$\mathcal{H} = H - i\hbar\frac{\partial}{\partial t}. \tag{6.38}$$

The eigenvalue ϵ is called a quasi-energy and the solution of Eq. (6.37) is a quasi-energy state or Floquet state. For each eigenstate ϕ_ϵ of $\mathcal{H}$ with eigenvalue ϵ there is a family of eigenstates $\phi_\epsilon e^{ik\Omega t}$ which belong to the same Floquet state. The states have eigenvalues $\epsilon + k\hbar\Omega$ and k takes integer values.

Floquet states form a convenient basis for modelling a periodically-driven system; the time-dependence of the drive is transferred onto the basis states, such that the coupling between Floquet states is time-independent. This means methods for solving the time-independent Schrödinger equation or Liouville equation may be used. The downside is that the dimension of a system is larger in the Floquet basis than in the bare state basis [13].

The Floquet basis may be used to gain insight into a periodically-driven system. In Fig. 6.7 a manifold of Rydberg $nD_{3/2}$ states is expanded in the Floquet basis and the coupling of Eq. (6.32) is represented. The difference in quasi-energy between coupled levels is given by both $\hbar\Omega_{\mathrm{rf}}$ and the Zeeman splitting of the levels. As a

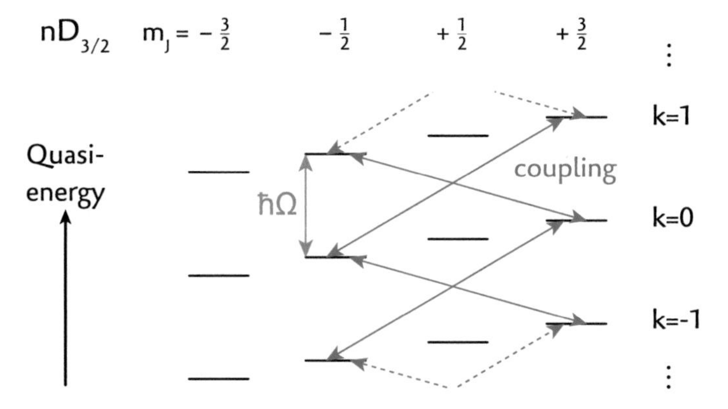

Fig. 6.7 Quadrupole field-driven coupling between levels shown in the Floquet basis: k is an index which labels Floquet modes, states with $\Delta m_J = 2$ and $\Delta k = 1$ are coupled. For clarity only coupling between $m_J = -\frac{1}{2}$ and $\frac{3}{2}$ states is shown

result the excitation spectra of Rydberg $D_{3/2}$-states include Floquet sidebands which are positioned at $\pm\Omega_{\text{rf}}$ relative to next-neighbouring Zeeman sublevels ($\Delta m_J = \pm 2$).

6.2.3 *Experimental Results*

The coupling between Rydberg states driven by the oscillating quadrupole field is observed in the excitation spectrum of $27D_{3/2}$. The ion is prepared in a mixture of $4D_{3/2}$ sublevels. The data shown in Fig. 6.8a is recorded using UV laser light which drives the two-photon transitions $4D_{3/2},\ m_J = \frac{3}{2} \leftrightarrow 27D_{3/2},\ m_J = \frac{3}{2}$ and $4D_{3/2},\ m_J = \frac{1}{2} \leftrightarrow 27D_{3/2},\ m_J = \frac{1}{2}$. These transitions are selected by using appropriate polarisations of the UV laser beams (see Sect. 4.5). The excitation spectra corresponding to these two transitions overlap, this is because neither transition is affected by the Zeeman effect, and because both Rydberg states couple to the next-neighbouring sublevel with $\Delta m_J = -2$.

Spectra are measured using two different strengths of trapping fields; Floquet sidebands are observed in both cases. When stronger trapping fields are used the coupling C between the levels is higher and the sidebands are more pronounced. The sidebands appear near to multiples of the frequency of the oscillating trapping electric field $\Omega_{\text{rf}} = 2\pi \times 18.2\,\text{MHz}$.

H_{rf} couples states with $\Delta m_J = \pm 2$. As shown in Fig. 6.8b, the first-order sidebands are offset from the next-neighbouring Zeeman levels by $\pm\Omega_{\text{rf}}$ and thus the offset from the carrier transition is $(\pm\Omega_{\text{rf}} - 2\pi \times 6.5\,\text{MHz})$ in the magnetic field of strength $B = 0.29\,\text{mT}$. The second-order sidebands are offset from the carrier transition by $\pm 2\Omega_{\text{rf}}$.

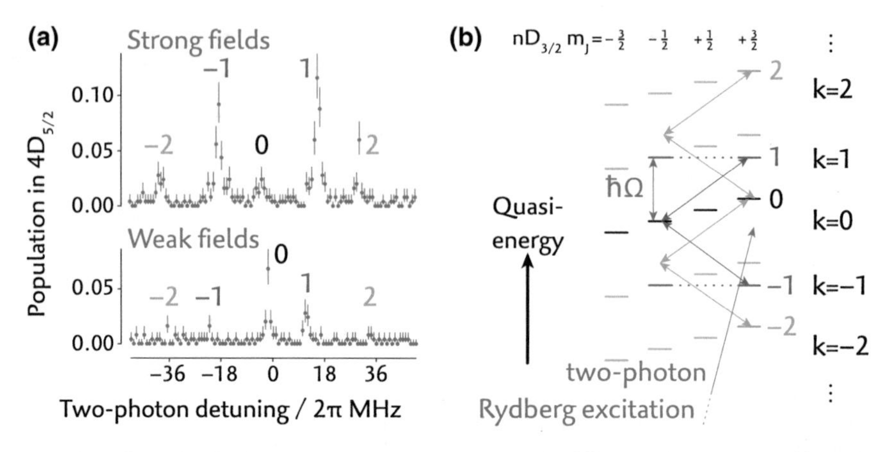

Fig. 6.8 **a** The oscillating quadrupole trapping field couples the $27D_{3/2}$ sublevels, this results in Floquet sidebands in $27D_{3/2}$ excitation spectra. The trapping field oscillates at $\Omega_{\text{rf}} = 2\pi \times 18.2\,\text{MHz}$; first-order Floquet sidebands are observed around $2\pi \times 18\,\text{MHz}$ from the carrier resonance, second-order sidebands are observed around $2\pi \times 36\,\text{MHz}$ from the carrier. The sidebands are stronger when stronger trapping fields are used. The frequency axis has uncertainty $\approx 2\pi \times 2\,\text{MHz}$ in the centre frequency and $\sim$10% uncertainty in a frequency range, as explained in the body text. Error bars indicate quantum projection noise (68% confidence interval). **b** The relevant trap-induced coupling for state $nD_{3/2}$, $m_J = \frac{3}{2}$ is shown in the Floquet basis. First-order Floquet sidebands (red) result from coupling to the $m_J = -\frac{1}{2}$ sublevel. Second-order sidebands (green) result from coupling first to the $m_J = -\frac{1}{2}$ sublevel then to the $m_J = \frac{3}{2}$ sublevel

Using the theory value of the quadrupole moment $\Theta(27D_{3/2}) = 5.3 \times 10^4\, e\, a_0{}^2$ (see Sect. 2.2) and Eqs. (6.31), (6.33) we can estimate the coupling strength C and the energy shift $|E_s|$: With relatively strong trapping fields (field gradients $\alpha = 8.2 \times 10^8\,\text{V}\,\text{m}^{-2}$ and $\beta = 6.8 \times 10^6\,\text{V}\,\text{m}^{-2}$) $C = 2\pi \times 34\,\text{MHz}$ and $|E_s|/\hbar = 2\pi \times 490\,\text{kHz}$. With weaker trapping fields ($\alpha = 3.3 \times 10^8\,\text{V}\,\text{m}^{-2}$ and $\beta = 5.7 \times 10^5\,\text{V}\,\text{m}^{-2}$) $C = 2\pi \times 14\,\text{MHz}$ and $|E_s|/\hbar = 2\pi \times 41\,\text{kHz}$. Floquet sidebands emerge because $C \not\ll \Omega_{\text{rf}}$. The energy shifts $|E_s|$ are not resolved.

The frequency axis in Fig. 6.8a has uncertainty $\approx 2\pi \times 2\,\text{MHz}$ in the centre frequency and $\sim$10% uncertainty in a frequency range. These uncertainties result because of the method used to scan the 307 nm laser frequency: The frequency was scanned by changing the length of the 615 nm laser frequency reference using piezoelectric rings (Sect. 3.3.4); the piezoelectric rings respond linearly only to a first approximation, they also show hysteresis effects. The centre frequency is estimated by measuring excitation spectra between other sublevels of $4D_{3/2}$ and $27D_{3/2}$ using different UV laser light polarisations, then taking the average centre frequency of all the excitation spectra. Further data has been collected since introduction of the AOM used to scan the 307 nm laser frequency. This data is published in [3] and shows excitation spectra of $24D_{3/2}$. The spectra are, however, less easy to interpret because of ac-Stark shifts.

References

1. Müller M, Liang L, Lesanovsky I, Zoller P (2008) Trapped Rydberg ions: from spin chains to fast quantum gates. New J Phys 10:093009
2. Facon A et al (2016) A sensitive electrometer based on a Rydberg atom in a Schrödinger-cat state. Nature 535:262–265
3. Higgins G et al (2017) Single strontium Rydberg ion confined in a Paul trap. Phys Rev X^7:021038
4. Higgins G, Pokorny F, Zhang C, Hennrich M (2019) Highly-polarizable ion in a Paul trap. arXiv:1904.08099
5. Li W, Lesanovsky I (2014) Entangling quantum gate in trapped ions via Rydberg blockade. Appl Phys B 114:37–44
6. Berkeland DJ, Miller JD, Bergquist JC, Itano WM, Wineland DJ (1998) Minimization of ion micromotion in a Paul trap. J Appl Phys 83:5025–5033
7. Keller J, Partner HL, Burgermeister T, Mehlstäubler TE (2015) Precise determination of micromotion for trapped-ion optical clocks. J Appl Phys 118:104501
8. Itano WM (2000) External-field shifts of the $^{199}Hg^+$ optical frequency standard. J Res Natl Inst Stand Technol 105
9. Shaniv R, Akerman N, Ozeri R (2016) Atomic quadrupole moment measurement using dynamic decoupling. Phys Rev Lett 116:140801
10. Jiang D, Arora B, Safronova MS (2008) Electric quadrupole moments of metastable states of Ca^+, Sr^+, and Ba^+. Phys Rev A 78:022514
11. Chwalla M (2009) Precision spectroscopy with 4^oCa^+ ions in a Paul trap. PhD thesis, Universität Innsbruck
12. Friedrich H (2017) Theoretical atomic physics. Springer International Publishing, New York
13. Bain AD, Dumont RS (2001) Introduction to Floquet theory: the calculation of spinning sideband intensities in magic-angle spinning NMR. Concepts Magn Reson 13:159–170

Chapter 7
Coherent Excitation of Rydberg States

In neutral Rydberg atom systems a host of quantum mechanical phenomena have been investigated, several of which are reviewed in Sect. 1.2. Quantum mechanical phenomena are observed most easily in isolated systems with strong coupling between only a few levels. With our understanding of Rydberg ion trap effects (see Chap. 6) we isolate a single Rydberg level and couple it with two other atomic levels using light fields; we then investigate the quantum-mechanical phenomena which appear.

While coupling between two levels gives rise to Rabi oscillations, coupling between the levels of a three-level system results in a range of phenomena [1], including the Autler–Townes effect (reported in Sect. 7.2), electromagnetically-induced transparency (EIT), two-photon Rabi oscillations (Sect. 7.3), as well as stimulated Raman adiabatic passage (STIRAP) (Sect. 7.4). Using STIRAP for coherent Rydberg excitation and deexcitation we measure the Rydberg state lifetime (Sect. 7.4.2) and carry out a single-qubit Rydberg phase gate (Sect. 7.4.3). This gate demonstrates the basic workings of a Rydberg ion quantum computer. Many of the results in this chapter are published in Physical Review Letters [2] or else are under review—a working manuscript is available at [3].

Before describing the results, the three-level system is introduced in Sect. 7.1.

7.1 The Three-Level System

A single Rydberg level is coupled to two other atomic levels using two UV laser fields. To isolate a single Rydberg level we avoid trap-driven coupling between Rydberg states (Sect. 6.2) by using a Rydberg S-state and we mitigate effects of the trap on the highly-polarisable Rydberg state (Sect. 6.1) by employing radial sideband cooling and by minimising micromotion.

G. Higgins, *A Single Trapped Rydberg Ion*, Springer Theses,
https://doi.org/10.1007/978-3-030-33770-4_7

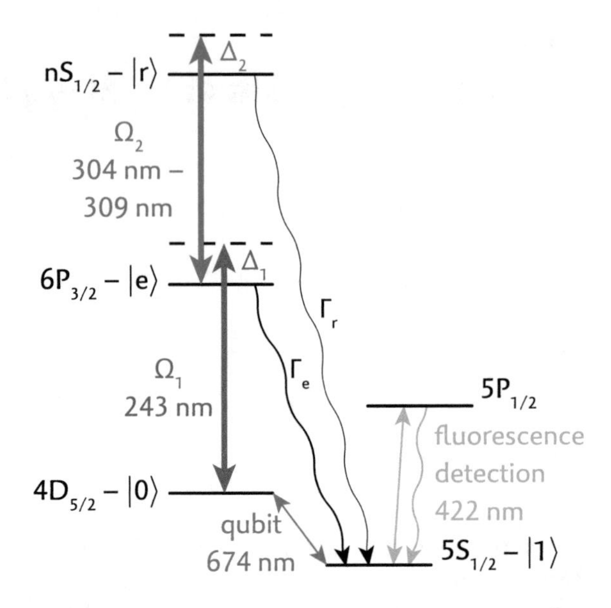

Fig. 7.1 Three atomic levels of $^{88}\mathrm{Sr}^+$ are coupled by two UV laser fields. Population in $|e\rangle$ or $|r\rangle$ decays mostly to $5S_{1/2}$; detection of scattered 422 nm light heralds excitation from $|0\rangle$ and decay to $5S_{1/2}$. A $4D_{5/2}$ sublevel and a $5S_{1/2}$ sublevel are used to store a qubit

The three levels coupled by light fields are $|0\rangle \equiv 4D_{5/2},\ m_J = -\frac{5}{2}$, $|e\rangle \equiv 6P_{3/2}$, $m_J = -\frac{3}{2}$ and $|r\rangle \equiv nS_{1/2},\ m_J = -\frac{1}{2}$. A sublevel of $4D_{5/2}$ is used rather than a sublevel of $4D_{3/2}$ since population may be initialised in a single $4D_{5/2}$ sublevel (Sect. 3.1.3). The levels are in a ladder configuration, as shown in Fig. 7.1.

Within the rotating wave approximation the coupling Hamiltonian describing the light-matter interaction is given by

$$H_c = \frac{\hbar}{2}\begin{pmatrix} 0 & \Omega_1 & 0 \\ \Omega_1 & 2\Delta_1 & \Omega_2 e^{i\phi} \\ 0 & \Omega_2 e^{-i\phi} & 2\Delta_1 + 2\Delta_2 \end{pmatrix} \tag{7.1}$$

in the basis $\{|0\rangle, |e\rangle, |r\rangle\}$. Ω_1, Ω_2, Δ_1 and Δ_2 are the Rabi frequencies and detunings from resonance of the first and second Rydberg excitation laser fields, and ϕ is the relative phase of the Rydberg excitation laser fields within the rotating frame. The lifetimes and decay rates of $|e\rangle$ and $|r\rangle$ are given by τ_e, Γ_e and τ_r, Γ_r.

7.2 Autler–Townes Effect

When two levels of a three-level system are coupled by an intense field, the absorption peak of a transition involving the third level, probed by a weak field, is split [4]. This phenomenon is called the Autler–Townes effect.

The first Rydberg excitation laser field probes the $|0\rangle \leftrightarrow |e\rangle$ transition while the second Rydberg excitation laser field couples $|e\rangle$ and $|r\rangle$. The $|0\rangle \leftrightarrow |e\rangle$ resonance lineshape is observed by scanning the frequency of the probe laser across the resonance, with $\Omega_1 \ll \Gamma_e$, as is described in Sect. 4.1.1. When the $|e\rangle$ and $|r\rangle$ levels are strongly coupled ($\Omega_2 > \Gamma_e,\ \Gamma_r$), the absorption profile of the probe field shows two peaks. When the coupling field is resonant with the $|e\rangle \leftrightarrow |r\rangle$ transition ($\Delta_2 = 0$) the splitting of the peaks is given by the coupling field Rabi frequency Ω_2. This effect is shown in Fig. 7.2.

When $\Omega_2 > \Gamma_e \gg \Omega_1$ the eigenstates and eigenvalues of Eq. (7.1) are

$$|\phi_0\rangle = |0\rangle \qquad E_0 = 0 \tag{7.2}$$

$$|\phi_\pm\rangle = \frac{-\Delta_2 \pm \sqrt{\Delta_2^2 + \Omega_2^2}}{\Omega_2}|e\rangle + |r\rangle \qquad E_\pm = \frac{\hbar}{2}\left(\Delta_2 \pm \sqrt{\Delta_2^2 + \Omega_2^2}\right). \tag{7.3}$$

$|\phi_0\rangle$ is the bare atomic state $|0\rangle$, $|\phi_\pm\rangle$ are light-dressed states. When $\Delta_2 = 0$ the dressed eigenstates simplify to $|\phi_\pm\rangle = |e\rangle \pm |r\rangle$ and the difference between the eigenenergies becomes $E_+ - E_- = \hbar\Omega_2$. Thus the resonance peaks in Fig. 7.2 correspond to excitation of the dressed states $|e\rangle \pm |r\rangle$, split by Ω_2. When $|\Delta_2| \gg \Omega_2$ the eigenenergies $E_\pm$ take the familiar form of the ac-Stark shift.

By fitting a model of the resonance structure to the experimental data we can extract Ω_2 and Δ_2. The model is derived by including decay channels from $|e\rangle$ and $|r\rangle$ to the $5S_{1/2}$ state. The derivation involves adiabatic elimination of $|e\rangle$ and $|r\rangle$, which is justified as follows: The decay rates of the dressed states are of similar magnitudes as the decay rate of $|e\rangle$, which greatly exceeds Ω_1; $\Gamma_e \gg \Omega_1$. This means

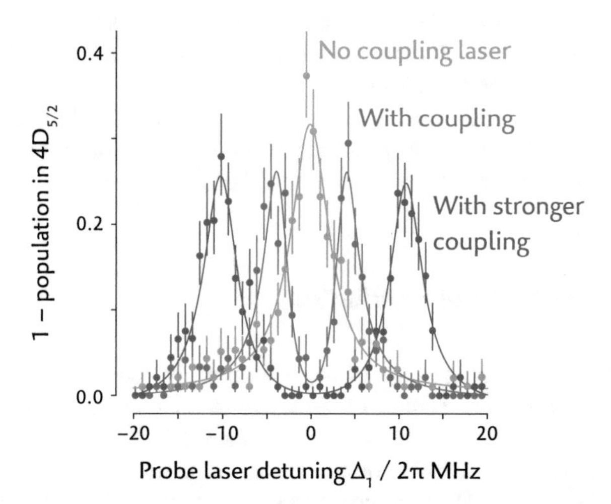

Fig. 7.2 Autler–Townes splitting in the three-level system. The ion is prepared in $|0\rangle$ and illuminated by the probe laser light. The Lorentzian absorption profile of $|0\rangle \leftrightarrow |e\rangle$ is observed when the coupling laser field is switched off. When the resonant coupling field is switched on an Autler–Townes doublet emerges, with splitting Ω_2. The solid lines are fits to the data; the fits return $\Omega_2 = 2\pi \times (7.88 \pm 0.25)$ MHz for the blue data points and $\Omega_2 = 2\pi \times (20.9 \pm 0.4)$ MHz for the dark blue data points. Here $|r\rangle = 42S_{1/2},\ m_J = -\frac{1}{2}$. Error bars indicate quantum projection noise (68% confidence interval)

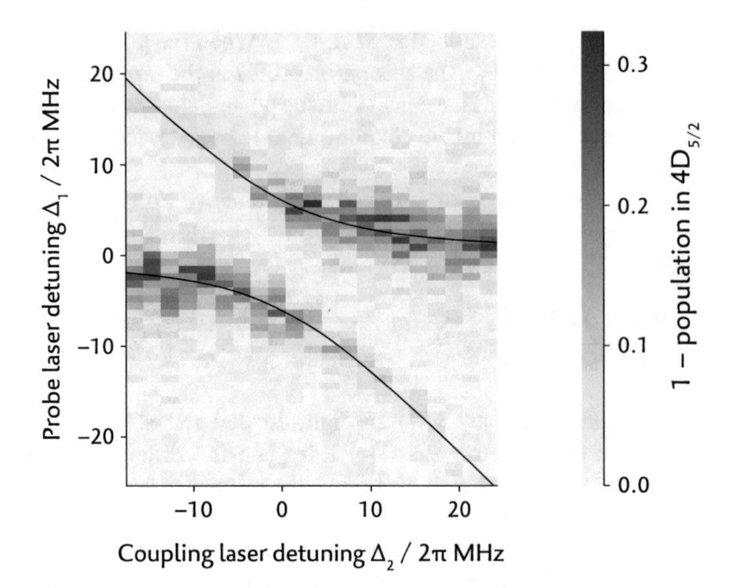

Fig. 7.3 Avoided crossing of the light-dressed states eigenenergies, described by Eq. (7.3). Resonance positions follow the ac-Stark formula $\Delta_1 = \frac{1}{2}\left(-\Delta_2 \pm \sqrt{\Delta_2{}^2 + \Omega_2{}^2}\right)$, which is shown by the black lines. A fit to the data returns $\Omega_2 = 2\pi \times (12.08 \pm 0.08)$ MHz. Here $|r\rangle = 42S_{1/2},\ m_J = -\frac{1}{2}$

that the dressed states do not become significantly populated, and thus the levels $|e\rangle$ and $|r\rangle$ do not become significantly populated. As is shown in Fig. 7.2 the model describes the experimental data well.

The measurement of the resonance structure is repeated as Δ_2 is varied; results are shown in Fig. 7.3. The resonance positions follow the ac-Stark formula. The results show the avoided crossing of the dressed states' eigenenergies $E_{\pm}$.

7.3 Two-Photon Rabi Oscillations

The theory proposals for systems of trapped Rydberg ions in Sect. 1.3.1 involve coherent excitation and deexcitation of Rydberg states. Coherent excitation and deexcitation is demonstrated in this section by two-photon Rabi oscillations. In Sect. 7.4 coherent excitation and deexcitation is carried out using stimulated Raman adiabatic passage (STIRAP). The two approaches use notably different laser parameters, for instance, synchronous UV laser pulses are used in this section, while STIRAP involves a sequence of laser pulses.

This section proceeds as follows: Experimental requirements are explained in terms of the theory behind two-photon Rabi oscillations in Sect. 7.3.1. Experimental results are discussed in Sect. 7.3.2.

7.3.1 Theoretical Background

This method results in coherent transfer of population between $|0\rangle$ and $|r\rangle$ without the lossy state $|e\rangle$ becoming populated. To avoid populating $|e\rangle$ the Rabi frequencies and detunings satisfy $\Omega_1 \ll |\Delta_1|$ and $\Omega_2 \ll |\Delta_2|$. Two-photon Rabi oscillations also require that the two-photon detuning between $|0\rangle$ and $|r\rangle$ $\Delta_{\text{2-photon}} \approx \Delta_1 + \Delta_2$ is small compared with the effective two-photon Rabi frequency Ω_{eff}.

Population is unlikely to be transferred to $|e\rangle$. Because of fast radiative decay out of the lossy state $|e\rangle$, with rate Γ_e, significant population cannot build up in $|e\rangle$. We can assume the rate of change of population in $|e\rangle$ is zero and adiabatically eliminate this state. The coupling Hamiltonian H_c [Eq. (7.1)] then becomes

$$H_c' = \frac{\hbar}{2}\begin{pmatrix} -\frac{\Omega_1^2}{2\Delta_1} & -\frac{\Omega_1\Omega_2}{2\Delta_1} \\ -\frac{\Omega_1\Omega_2}{2\Delta_1} & -\frac{\Omega_2^2}{2\Delta_1} + 2\Delta_1 + 2\Delta_2 \end{pmatrix} \tag{7.4}$$

in the basis $\{|0\rangle, |r\rangle\}$. The condition for two-photon resonance comes from the diagonal elements

$$\Delta_{\text{2-photon}} = \Delta_1 + \Delta_2 + \underbrace{\frac{\Omega_1^2 - \Omega_2^2}{4\Delta_1}}_{\text{ac-Stark shifts}} = 0. \tag{7.5}$$

This condition already accounts for level shifts which result from the ac-Stark effect.[1]

The off-diagonal elements present the coupling between $|0\rangle$ and $|r\rangle$ with effective Rabi frequency

$$\Omega_{\text{eff}} = \frac{\Omega_1\Omega_2}{2\Delta_1}. \tag{7.6}$$

Unwanted off-resonant excitation of $|e\rangle$ from $|0\rangle$ and from $|r\rangle$ occurs with rates [5]

$$\begin{aligned} R_{0\to e} &= \frac{\Gamma_e}{4}\frac{\Omega_1^2}{\Delta_1^2 + \frac{\Omega_1^2}{2} + \frac{\Gamma_e^2}{4}} \approx \frac{\Gamma_e\Omega_1^2}{4\Delta_1^2}, \\ R_{r\to e} &= \frac{\Gamma_e}{4}\frac{\Omega_2^2}{\Delta_2^2 + \frac{\Omega_2^2}{2} + \frac{\Gamma_e^2}{4}} \approx \frac{\Gamma_e\Omega_2^2}{4\Delta_2^2}. \end{aligned} \tag{7.7}$$

Efficient transfer of population between $|0\rangle$ and $|r\rangle$ requires $\Omega_{\text{eff}} \gg R_{0\to e}$, $R_{r\to e}$. The ratios of Ω_{eff} to $R_{0\to e}$ and to $R_{r\to e}$ can be increased by increasing the magnitudes of the detunings Δ_1 and Δ_2.

There is, however, a detrimental effect of increasing $|\Delta_1|$; Ω_{eff} depends inversely on $|\Delta_1|$ [from Eq. (7.6)] and if Ω_{eff} is too small then Rydberg state decay Γ_r limits

[1] Interestingly these shifts cancel when $\Omega_1 = \Omega_2$ for a three-level ladder system, provided that coupling to levels outside the three-level system can be ignored.

the transfer efficiency. And thus when choosing Δ_1 and Δ_2 a compromise must be reached between losses due to scattering off $|e\rangle$ and losses due to Rydberg state decay.

Finally laser linewidths Δ_{laser} contribute to the two-photon detuning and must be small compared with the two-photon Rabi frequency for efficient transfer between $|0\rangle$ and $|r\rangle$.

In summary, to achieve high-visibility two-photon Rabi oscillations we require

$$\Omega_{\text{eff}} \gg \Gamma_r,\ R_{0\to e},\ R_{r\to e},\ \Delta_{\text{2-photon}},\ \Delta_{\text{laser}}. \tag{7.8}$$

7.3.2 Experimental Results

Two-photon Rabi oscillations between $|0\rangle$ and $|r\rangle = 46S_{1/2},\ m_J = -\frac{1}{2}$ are observed for a sideband cooled ion, but not for a Doppler cooled ion, as shown in Fig. 7.4.

We use $\Omega_{\text{eff}} = 2\pi \times (1.23 \pm 0.09)\,\text{MHz} \gg \Gamma_{46S} = 2\pi \times 34.4\,\text{kHz}$ (calculated using wavefunctions described in Sect. 2.2), $R_{0\to e} = (15.4 \pm 2.0)\,\mu\text{s}^{-1}$, $R_{r\to e} = (88 \pm 8)\,\mu\text{s}^{-1}$ and $\Delta_{\text{laser}} \approx 2\pi \times 100\,\text{kHz}$ (for each of the UV lasers—see Sect. 3.3.4).

The two-photon detuning $\Delta_{\text{2-photon}}$ depends on the number of radial phonons, due to effects of the additional trapping potential experienced by the highly-polarisable Rydberg ion (Sect. 6.1). After radial sideband cooling, an ion has narrow radial

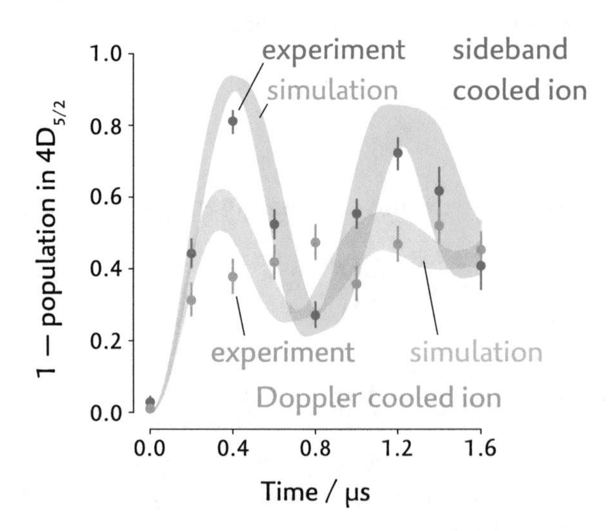

Fig. 7.4 Two-photon Rabi oscillations between $|0\rangle$ and $|r\rangle$ are observed with a radial sideband cooled ion, and simulation results are consistent with experimental results. Rabi oscillations are not observed with a Doppler cooled ion, likely due to trap effects. Simulation results do not agree as well with the Doppler cooled ion experimental results. Error bars indicate quantum projection noise (68% confidence interval). The shaded areas shows the central 68% simulation results calculated at each time step

phonon distributions and Rydberg resonances are narrow. The two-photon detuning is then $\Delta_{2\text{-photon}} = 2\pi \times (0 \pm 150)$ kHz $\ll \Omega_{\text{eff}}$. The criteria for two-photon Rabi oscillations [Eq. (7.8)] are satisfied and the Rabi oscillations are shown in Fig. 7.4.

The Doppler cooled ion has broader radial phonon distributions and the two-photon detuning depends on the number of radial phonons in each experimental run. The mean absolute detuning is not much smaller than the effective Rabi frequency $|\Delta_{2\text{-photon}}| \approx 2\pi \times 600$ kHz $\not\ll \Omega_{\text{eff}}$. This goes some way to explaining why Rabi oscillations are not observed for the Doppler cooled ion.

Some of the parameters listed thus far are based on experimentally-determined parameters: $\Gamma_e = 2\pi \times (4.9 \pm 0.4)$ MHz, $\Omega_1 = 2\pi \times (23.0 \pm 1.5)$ MHz and $\Delta_1 = 2\pi \times (515 \pm 15)$ MHz (found using the method in Sect. 4.1.1), and $\Omega_2 = 2\pi \times (55 \pm 1)$ MHz (method in Sect. 7.2). Regarding Δ_2: the two-photon resonance condition [Eq. (7.5)] is met by varying the frequency of the second Rydberg excitation laser until the Rabi oscillation visibility is maximised. After radial sideband cooling the average numbers of radial phonons are estimated to be $\bar{n}_x \approx \bar{n}_y \approx 0.2$, while after Doppler cooling $\bar{n}_x \approx 10.8 \pm 3.0$ and $\bar{n}_y \approx 15.6 \pm 3.0$ (see Sect. 3.1.1). The radial trapping frequencies are altered for the Rydberg ion by $\Delta\omega_x \approx \Delta\omega_y \approx -2\pi \times 40.5$ kHz (see Sect. 6.1).

Simulations of the Experiment

To find what limits the Rabi oscillation visibility, the experiment is simulated by numerically solving the Lindblad master equation using the Python framework QuTiP [6]. Experimentally-determined parameters are used.

To simulate the sideband cooled ion, the simulation includes four electronic levels $\{|0\rangle, |e\rangle, |r\rangle, 5S_{1/2}\}$ and eight dimensions which represent the motional degrees of freedom in the radial directions; population is assumed to be spread over the motional degrees of freedom according to thermal distributions with an average of 0.2 phonons in each mode. The change of the trapping potential experienced by Rydberg ions is included by a phonon-number-dependent resonance shift and by phonon-number-changing transitions coupled with the Rydberg excitation. The strengths of the phonon-number-changing transitions are determined by Franck–Condon factors, and the contribution from any mismatch between the positions of the nulls of the two quadrupole trapping fields is neglected (see Sect. 6.1.2).

Uncertainties in the experimental parameters are accounted for using a Monte Carlo method: parameters are drawn from Gaussian probability distributions (the widths of the distributions are given by the uncertainties of the experimental parameters) and the simulation is repeated 500 times. From the set of simulations the 68% confidence interval for the $|0\rangle$ population at each simulation step is determined and plotted as the shaded area in Fig. 7.4. The simulation results match the experimental results well.

Higher visibility Rabi oscillations between low-lying states and Rydberg states have been carried out in neutral Rydberg atom experiments [7]. Further simulations indicate the oscillation visibility in our experiment is mainly limited by the Rydberg excitation lasers linewidths as well as Rydberg state decay. The visibility may be improved by exciting higher Rydberg states with longer lifetimes, by reducing the linewidths of the UV lasers and by using higher effective two-photon Rabi frequencies.

Trap effects related to the high Rydberg state polarisability (Sect. 6.1) are effectively mitigated by radial sideband cooling; simulations which do not account for trap effects return oscillations with only marginally higher visibilities (0.1% higher) than simulations which include trap effects.

The experiment with a Doppler cooled ion is also simulated using experimentally-determined parameters and a Monte Carlo method. The Doppler cooled ion has population spread over many more phonon number states than the sideband cooled ion; to speed up the Monte Carlo simulations phonon-number-changing transitions are not included. This is justified because simulations indicate that the effect of phonon-number-changing transitions on the oscillation visibility is dwarfed by the effect of the phonon-number-dependent resonance shift, provided that the positions of the nulls of the two quadrupole trapping fields overlap (Sect. 6.1.1).

The simulations predict low-visibility Rabi oscillations, which are not observed experimentally. It may be that the nulls of the quadrupole trapping fields are significantly separated, and that strongly-driven phonon-number-changing transitions cause dephasing which dampens the oscillations such that oscillations are not resolved (Sect. 6.1.2). Such phonon-number-changing transitions are stronger for high phonon number states than for low phonon number states (see Fig. 6.3). This means the phonon-number-changing transitions may be negligible for a sideband cooled ion while they may significantly decrease the oscillation visibility for a Doppler cooled ion.

7.4 Stimulated Raman Adiabatic Passage

In addition to using synchronous UV laser pulses (as in Sect. 7.3), the $|0\rangle \leftrightarrow |r\rangle$ transition can be driven coherently using a sequence of pulses in a technique called stimulated Raman adiabatic passage (STIRAP).

STIRAP is a method for transferring population between two quantum states; the transfer proceeds by using light fields to couple the two states to a third intermediate state. When implemented correctly, population is transferred between the two states without the intermediate state becoming populated. And thus the transfer process does not suffer from losses due to spontaneous emission from the intermediate state, even when the STIRAP transfer occurs over timescales longer than the intermediate state lifetime. STIRAP has the additional advantage that it is insensitive to small variations of experimental parameters including laser powers, laser frequencies, pulse timing and pulse shapes [8].

In Sect. 7.4.1 the method is described further and population transfer between $|0\rangle$ and $|r\rangle$ with $(91 \pm 3)\%$ efficiency is shown. In Sect. 7.4.2 STIRAP is used to measure a Rydberg state lifetime. Finally in Sect. 7.4.3 geometric phases are introduced to a qubit state via Rydberg excitation and deexcitation by STIRAP and a single-qubit geometric gate is carried out. This gate demonstrates the basic operation of a Rydberg ion quantum computer.

7.4.1 Coherent Rydberg Excitation Using STIRAP

When the UV laser detunings cancel each other ($\Delta_1 + \Delta_2 = 0$) the coupling Hamiltonian [Eq. (7.1)] has a "dark" eigenstate

$$|\Phi_{\text{dark}}\rangle = \Omega_2 e^{i\phi}|0\rangle - \Omega_1|r\rangle. \tag{7.9}$$

The dark state is named so because it does not contain any component of $|e\rangle$ and thus the Rydberg excitation laser light is not scattered by an ion in the dark state during timescales much less than the Rydberg state lifetime τ_r.

Population transfer by STIRAP proceeds by having the state vector follow the dark state as the ratio $\frac{\Omega_1}{\Omega_2}$ is varied and the character of the dark state is changed between $|0\rangle$ and $|r\rangle$. The state vector follows the dark state provided the dark state changes adiabatically. The adiabaticity criterion [9] is

$$|\dot{\theta}(t)| \ll \sqrt{\Omega_1(t)^2 + \Omega_2(t)^2}, \tag{7.10}$$

where the mixing angle $\theta(t)$ is given by

$$\tan\theta(t) = \frac{\Omega_1(t)}{\Omega_2(t)}. \tag{7.11}$$

STIRAP Pulse Sequence

We implement STIRAP by varying Ω_1 and Ω_2 sinusoidally during the rise time t_{rise}. While $0 \le t \le t_{\text{rise}}$

$$\begin{aligned} \Omega_1(t) &= \Omega_{\text{peak}} \sin\frac{\pi t}{2t_{\text{rise}}}, \\ \Omega_2(t) &= \Omega_{\text{peak}} \cos\frac{\pi t}{2t_{\text{rise}}}, \end{aligned} \tag{7.12}$$

where the peak Rabi frequencies Ω_{peak} are the same for both Ω_1 and Ω_2. Pulse shaping is achieved by driving acousto-optic modulators (AOMs) using arbitrary waveform generators (Sect. 3.3.2). The shapes and the temporal overlap of the pulses are checked using photodiodes. Pulse sequences are shown in Fig. 7.5.

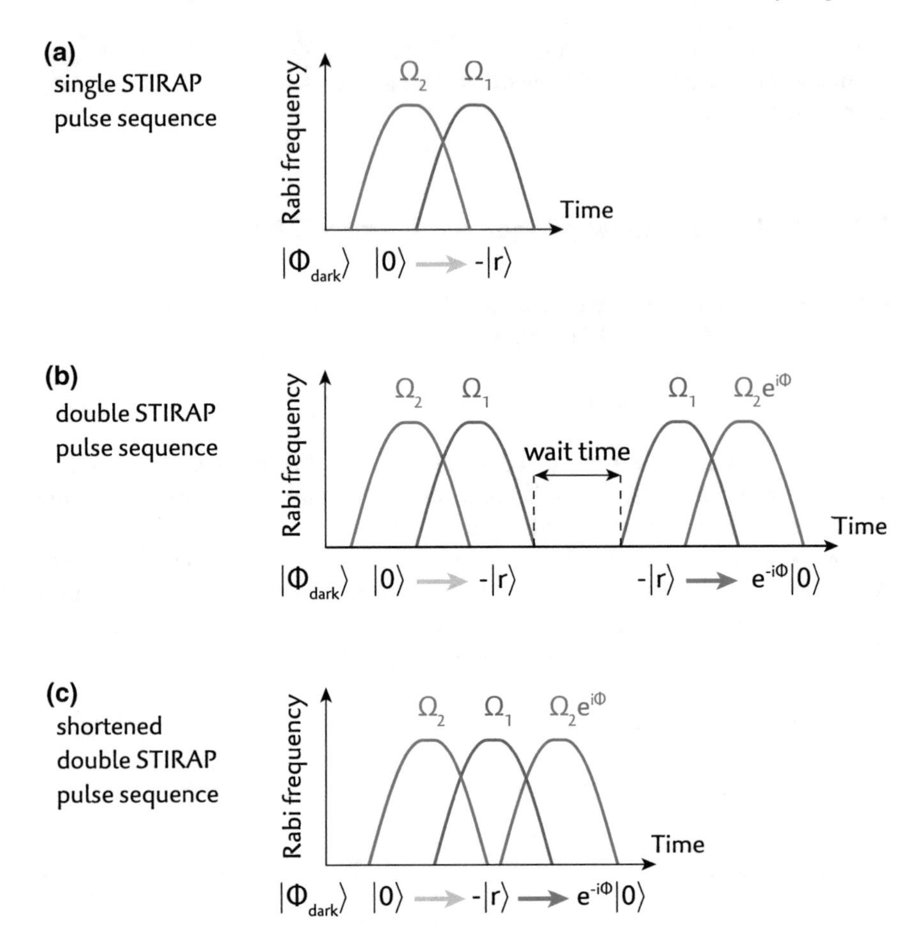

Fig. 7.5 STIRAP pulse sequences. The evolution of the dark state is written beneath each pulse sequence. The third sequence is a condensed version of the second sequence

The order of the pulses is counter-intuitive; to transfer population from $|0\rangle \rightarrow |r\rangle$ the 306 nm laser pulse is applied before the 243 nm laser pulse. Because the pulses are sinusoidally-shaped the mixing angle [Eq. (7.11)] changes at a constant rate during the rise time t_{rise}

$$\theta(t) = \frac{\pi t}{2t_{\text{rise}}} \tag{7.13}$$

and the adiabaticity criterion takes the simple form

$$\frac{\pi}{2t_{\text{rise}}} \ll \Omega_{\text{peak}}. \tag{7.14}$$

In the experiment the Rabi frequencies Ω_1, Ω_2 and the detunings Δ_1, Δ_2 are found using the methods in Sects. 4.1.1 and 7.2. The requirement $\Delta_1 + \Delta_2 = 0$ is met by setting $\Delta_1 = \Delta_2 = 0$. Ω_{peak} is limited by the maximum value of Ω_1 and thus by the 243 nm laser light intensity at the position of the ion. We use $t_{\text{rise}} = 200\,\text{ns}$ and $\Omega_{\text{peak}} \approx 2\pi \times 47\,\text{MHz}$ and satisfy the adiabaticity criterion $\frac{\pi}{2t_{\text{rise}}} = 2\pi \times 1.25\,\text{MHz} \ll \Omega_{\text{peak}} = 2\pi \times 47\,\text{MHz}$. t_{rise} must be long for the state vector to adiabatically follow the dark state, however if it is too long there are losses from Rydberg state decay ($\tau_r \approx 2.3\,\mu\text{s}$ for $42S_{1/2}$) and from decoherence due to finite laser linewidths. $t_{\text{rise}} = 200\,\text{ns}$ is chosen as a compromise. The Rydberg excitation lasers linewidths are each estimated to be $\sim 2\pi \times 100\,\text{kHz}$ (Sect. 3.3.4).

Detection of Population Transfer by STIRAP

State detection in our system takes longer than the lifetime of $|r\rangle$. As is described in Sects. 4.1.1 and 4.3, detection of population excited from $|0\rangle$ to $|e\rangle$ or from $|0\rangle$ to $|r\rangle$ relies on decay from $|e\rangle$ or $|r\rangle$ to $5S_{1/2}$ and $4D_{3/2}$; population in $5S_{1/2}$ and $4D_{3/2}$ is then distinguished from population in $|0\rangle$ by electron shelving (Sect. 3.1.4). This method does not distinguish successful transfer to $|r\rangle$ from scattering off $|e\rangle$; in both cases there is decay to $5S_{1/2}$ and $4D_{3/2}$ and thus this method is not able to confirm population transfer by STIRAP.

Instead population transfer is detected by comparing the population residing in $|0\rangle$ after application of the single STIRAP pulse sequence and application of the double STIRAP pulse sequence (pulse sequences are shown in Fig. 7.5). After the single STIRAP pulse sequence $(95^{+2}_{-5})\%$ of the population is removed from $|0\rangle$. After the double STIRAP pulse sequence up to 58% of the population resides in state $|0\rangle$. This indicates the second set of pulses in the double STIRAP pulse sequence returns population to $|0\rangle$ and thus demonstrates excitation and deexcitation by STIRAP. These results are shown in Fig. 7.6. The return of population to $|0\rangle$ is not perfect because the state vector does not follow the dark state perfectly, due to Rydberg state decay, finite laser linewidths and the short t_{rise}.

Using the shortened double STIRAP pulse sequence in Fig. 7.5c the detrimental effect of Rydberg state decay is reduced and $(83^{+5}_{-6})\%$ of the population is returned to $|0\rangle$. This indicates a transfer efficiency $\sqrt{(83^{+5}_{-6})\%} = (91 \pm 3)\%$, which exceeds the highest STIRAP efficiencies reported in systems of neutral Rydberg atoms (60%) [10–12]. In Sect. 7.4.3 ways to improve the transfer efficiency are discussed.

7.4.2 *Measurement of a Rydberg State Lifetime*

In this section the first lifetime measurement of a trapped Rydberg ion is described. This measurement relies upon coherent Rydberg excitation and deexcitation using STIRAP.

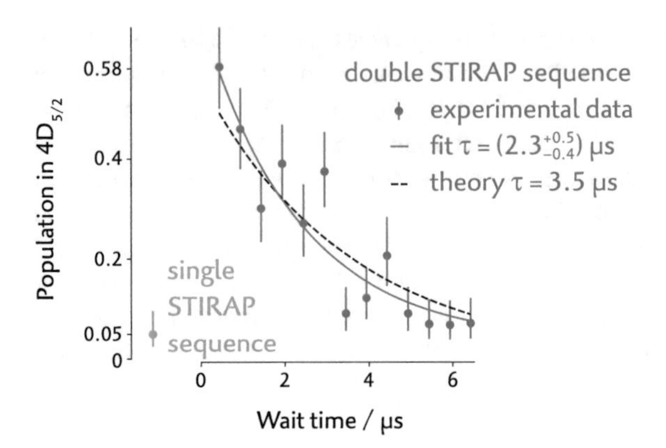

Fig. 7.6 Coherent Rydberg excitation by STIRAP shown by comparing the results of applying the single and the double STIRAP pulse sequences to an ion initially in state $|0\rangle$. $(5^{+5}_{-3})\%$ of the population remains in $|0\rangle$ after the single STIRAP pulse sequence; the rest of the population is transferred to $5S_{1/2}$ via $|e\rangle$ or $|r\rangle$. Most of the population may lie in $|0\rangle$ after the double STIRAP pulse sequence, this shows population is successfully excited to the Rydberg state $|r\rangle = 42S_{1/2},\ m_J = -\frac{1}{2}$ then de-excited to $|0\rangle$. The $42S_{1/2}$ state lifetime is obtained by varying the wait time between the two sets of STIRAP pulses and measuring the returned population. Error bars indicate quantum projection noise (68% confidence interval)

During the wait time between the two sets of pulses in the double STIRAP pulse sequence [see Fig. 7.5b] population may decay from $|r\rangle$; this population is not returned to $|0\rangle$ by the second set of STIRAP pulses. We determine the lifetime of $|r\rangle$ by measuring the population returned to $|0\rangle$ as the wait time is varied. We measure $\tau_{\text{exp}} = (2.3^{+0.5}_{-0.4})$ µs for $42S_{1/2}$, as shown in Fig. 7.6.

The theoretical value for the lifetime of $42S_{1/2}$ is $\tau_{300\,\text{K}} = 3.5$ µs when the surroundings have temperature 300 K (similar to the laboratory temperature). This value accounts for radiative decay as well as transitions driven by blackbody radiation (see Sect. 2.4).

The discrepancy between the theory value and the experimental value of the lifetime may be due to imperfect extinction of 306 nm laser light during the wait time of the double STIRAP pulse sequence. During this time weak 306 nm laser light could drive the $|r\rangle \leftrightarrow |e\rangle$ transition and population may be lost due to decay from $|e\rangle$. As a result the measured lifetime of $42S_{1/2}$ would be shortened.

Comparison between the theoretical value and the experimental value of the lifetime suggests there is no significant lifetime reduction because of confinement in the trap. This is an important result; if the Rydberg state lifetime was significantly shortened by the trap the viability of Rydberg ions as a quantum technology may be limited, since Rydberg state lifetimes place fundamental limits on gate fidelities [13] and resonance linewidths.

7.4.3 Geometric Phase Measurements

A geometric phase is introduced to qubit state $|0\rangle$ via Rydberg excitation and deexcitation using a STIRAP pulse sequence. This section proceeds as follows: the theory behind geometric phases is established, the experimental implementation is then described, the measurement of the geometric phase using a Ramsey-type experiment is presented, and finally a single-qubit Rydberg gate which uses the geometric phase is discussed.

Theory Behind Geometric Phases

When a quantum system is changed slowly in a cyclic fashion the eigenstates of the Hamiltonian accumulate dynamic and geometric phases [14]. The dynamic phase ϕ_{dyn} results from the evolution of an eigenenergy; the eigenstate $|n(t)\rangle$ with eigenenergy $\epsilon_n(t)$ which evolves during a cycle with period T accumulates dynamic phase

$$\phi_{\mathrm{dyn}} = -\frac{1}{\hbar}\int_0^T \epsilon_n(t')dt'. \tag{7.15}$$

The geometric phase accumulated is given by

$$\gamma = i\int_0^T \langle n\left(\vec{R}(t')\right)|\frac{d}{dt'}|n\left(\vec{R}(t')\right)\rangle dt' \tag{7.16}$$

$$= i\oint_{\mathcal{C}} \langle n\left(\vec{R}(t')\right)|\vec{\nabla}_{\vec{R}}n\left(\vec{R}(t')\right)\rangle\cdot d\vec{R}, \tag{7.17}$$

where the quantum system is parameterised by a set of parameters $\vec{R}$ and the system follows the closed path $\mathcal{C}$ in the parameter space such that $\vec{R}(T) = \vec{R}(0)$. Unlike the dynamic phase, the geometric phase does not depend on the rate at which the system evolves along $\mathcal{C}$.

Introduction of a Geometric Phase in the Experiment

In our system we accumulate a geometric phase during the shortened double STIRAP pulse sequence by varying the mixing angle θ [see Eq. (7.11)] and the phase difference between the UV laser fields within the rotating frame ϕ in a closed cycle, following a protocol recently used with a solid-state qubit [15]. The cycle is parameterised by $\vec{R} = \begin{pmatrix}\theta\\ \phi\end{pmatrix}$ and follows

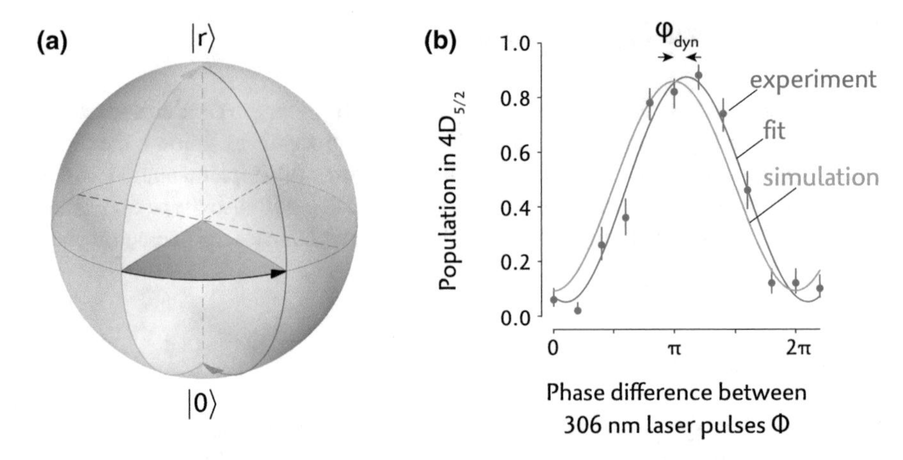

Fig. 7.7 Accumulation of a geometric phase during a Rydberg excitation process. **a** during the shortened double STIRAP pulse sequence the dark state moves in a 'tangerine-slice' trajectory on the surface of the Bloch sphere spanned by $|0\rangle$ and $|r\rangle$. The wedge angle is determined by the phase step of the 306 nm laser light. The mixing angle in Eq. (7.11) corresponds to the polar angle on the Bloch sphere. A geometric phase results from the curvature of the area enclosed by the trajectory. **b** The geometric phase is measured by using the qubit state $|1\rangle$ as a phase reference, and measuring the phase difference between $|0\rangle$ and $|1\rangle$ by a $\hat{\sigma}_y$-measurement. Here $|r\rangle = 42S_{1/2},\ m_J = -\frac{1}{2}$. Error bars indicate quantum projection noise (68% confidence interval)

$$\begin{pmatrix}0\\0\end{pmatrix} \to \begin{pmatrix}\pi\\0\end{pmatrix} \to \begin{pmatrix}\pi\\\Phi\end{pmatrix} \to \begin{pmatrix}0\\\Phi\end{pmatrix} \to \begin{pmatrix}0\\0\end{pmatrix}. \tag{7.18}$$

During the cycle the dark state traverses the surface of the Bloch sphere spanned by $|0\rangle$ and $|r\rangle$ in Fig. 7.7a; the mixing angle θ corresponds to the polar angle, the phase difference ϕ corresponds to the azimuthal angle.

The dark state moves from the 0-pole to the r-pole, then back to the 0-pole. The phase of the second Rydberg excitation laser light is changed by Φ when the dark state is at the r-pole, that is when $\Omega_2 = 0$, and the dark state returns to the 0-pole along a different meridian (line of constant ϕ). The complete path takes the form of a tangerine slice with wedge angle Φ.[2] The path circumscribes solid angle 2Φ and from Eqs. (7.9), (7.17) the geometric phase $-\Phi$ is accumulated throughout the process [15]. During the cycle the dark state changes from $|0\rangle \to -|r\rangle \to e^{-i\Phi}|0\rangle$.

Measurement of the Geometric Phase

While the global phase of a quantum state cannot be measured, phase differences between quantum states can be measured. The geometric phase introduced to qubit

[2]Note that ϕ represents the variable phase difference between the two UV laser fields within the rotating frame, while Φ represents the phase step introduced to the 306 nm laser light during the sequence and thus the wedge angle of the path followed by the dark state in Fig. 7.7a.

state $|0\rangle$ during the cycle is measured via a Ramsey experiment (which uses the qubit state $|1\rangle \equiv 5S_{1/2},\ m_J = -\frac{1}{2}$ as a phase reference) as follows:

1. The ion is prepared in superposition state $|\psi\rangle = |0\rangle + |1\rangle$ as described in Sect. 3.1.3.
2. The shortened double STIRAP pulse sequence is carried out and the state evolves to $|\psi\rangle = e^{-i\Phi}|0\rangle + |1\rangle$ (with perfect transfer efficiency).
3. A measurement in the $\hat{\sigma}_x$-basis carried out: the state is rotated about the $\hat{y}$ axis by $\frac{\pi}{2}$ to become $|\psi\rangle = -\sin\frac{\Phi}{2}|0\rangle + \cos\frac{\Phi}{2}|1\rangle$ and then electron shelving is used to distinguish population in the $\{|0\rangle, |1\rangle\}$ basis (Sect. 3.1.4). The fraction of the population projected onto $|0\rangle$ is $\sin^2\frac{\Phi}{2}$.

This Ramsey experiment is repeated as the laser phase step Φ is varied, the results are shown in Fig. 7.7b. Oscillatory behaviour is observed, confirming that a geometric phase is acquired during the shortened double STIRAP pulse sequence. The contrast of the oscillation $\mathcal{C} = (82 \pm 4)\%$ is less than unity because of the imperfect STIRAP transfer efficiency and decoherence due to finite laser linewidths. The average of the oscillation is less than 0.5 because population decays from $|r\rangle$ outside the $\{|0\rangle, |1\rangle\}$ manifold (mostly to $5S_{1/2},\ m_J = +\frac{1}{2}$).

This experiment is simulated by numerically solving the Lindblad master equation for the five-level system $\{|0\rangle, |e\rangle, |r\rangle, |1\rangle, 5S_{1/2},\ m_J = +\frac{1}{2}\}$ using experimentally-determined parameters with the Python framework QuTiP [6]. Excellent agreement is observed between the experimental and simulation results. Further simulations indicate the dynamic phase offset in the experimental data $\phi_{\text{dyn}} = (18 \pm 4)^\circ$ may be accounted for by a small two-photon detuning ($\sim 2\pi \times 100\,\text{kHz}$) from the $|0\rangle \leftrightarrow |r\rangle$ resonance and by the light shift from the second Rydberg excitation laser field on $|1\rangle$.

The simulation results show a lower contrast than the experimental results. Using the experimentally-determined parameters, simulations predict STIRAP transfer efficiency 90%. Repeating the simulations with Rydberg excitation laser linewidths $2\pi \times 64\,\text{kHz}$ the simulation contrast matches the experimental contrast and the simulations predict STIRAP transfer efficiency 91%. The good agreement between simulation results and experimental results indicate the simulations model our system well.

Simulations indicate the STIRAP transfer efficiency is limited by Rydberg state decay, Rydberg excitation laser linewidths and imperfect adiabatic following during the rise time t_{rise}. The transfer efficiency may be improved by using higher Rydberg-excitation laser light intensities, by exciting higher Rydberg states with longer lifetimes and by improving the frequency stabilisation of the Rydberg-excitation lasers.

Single-Qubit Rydberg Gate

The geometric phase accumulated during a cycle is insensitive to small changes in the path followed and thus geometric manipulation of quantum systems may allow noise-resilient quantum computation [16, 17].

We use the shortened STIRAP pulse sequence [Fig. 7.5c] to implement a single-qubit geometric phase gate. This gate is characterised using quantum process tomography for the case with phase step $\Phi = \pi$. Descriptions of quantum state tomography and quantum process tomography are available in [18, 19], brief descriptions of the techniques follow.

Quantum state tomography allows an unknown quantum state to be reconstructed. The same unknown state is prepared multiple times and measured using different measurement operators. The measurement operators must form a complete basis set in the Hilbert space of the system. In our system we measure a single qubit and use the Pauli operators $\{\hat{\sigma}_x, \hat{\sigma}_y, \hat{\sigma}_z\}$ as measurement operators, as described in Sect. 3.1.4. Measurements with the same measurement operator are typically repeated $\approx$50 times. Each measurement produces a binary outcome; measurements must be repeated for quantum states to be determined accurately.

With quantum process tomography a process is characterised by carrying out the process on known states then carrying out quantum state tomography on the post-process states. For a system with a d-dimensional Hilbert space copies of d^2 input states are required. Input states are chosen such that their density matrices form a basis set for the space of matrices. For our single-qubit system $d = 2$ and thus we prepare copies of the four input states $|0\rangle$, $|0\rangle + |1\rangle$, $|0\rangle + i|1\rangle$ and $|1\rangle$. State preparation is described in Sect. 3.1.3.

A process is conventionally described by a process matrix or χ-matrix. We use a maximum likelihood estimation in determining the χ-matrix. The reconstructed χ-matrix is shown in Fig. 7.8a. The ideal process is equivalent to a $\hat{\sigma}_z$ rotation. The

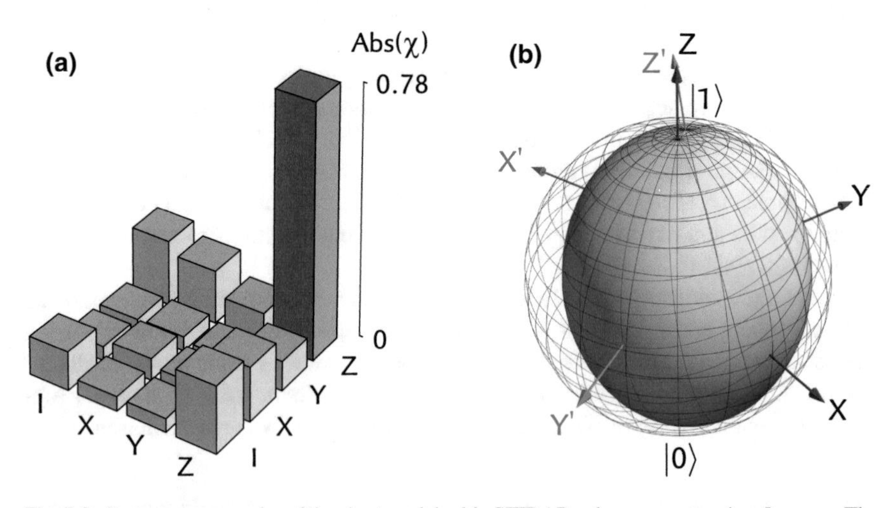

Fig. 7.8 Process tomography of the shortened double STIRAP pulse sequence using $\Phi = \pi$. **a** The absolute value of the reconstructed process matrix is shown. In the ideal process the ZZ-bar height is 1, the measured process fidelity is $(78 \pm 4)\%$. **b** The process matrix is visualised by comparing the pre-process (meshed) and post-process (blue) Bloch spheres: the Bloch sphere is rotated about the z-axis by $\approx\pi$ and inefficiencies cause the sphere to shrink. Here $|r\rangle = 42S_{1/2},\ m_J = -\frac{1}{2}$

process fidelity of $(78 \pm 4)\%$ is the overlap of the reconstructed χ-matrix with the χ-matrix for a $\hat{\sigma}_z$ rotation. The errors are estimated by a Monte Carlo method; due to quantum projection noise the measurement results have uncertainties described by binomial probability distributions. Samples are repeatedly drawn from the probability distributions and χ is repeatedly reconstructed. The process fidelities of the different χ-matrices are found and their distribution is used to estimate the uncertainty in the estimate of the process fidelity.

The reconstructed χ-matrix may be used to predict the effect of the process on any qubit state. This is represented in Fig. 7.8b; initial qubit states sitting on the meshed Bloch sphere are transformed by the process and the output states sit on the solid Bloch sphere. In the ideal process the solid Bloch sphere would be rotated about the z-axis by π; the imperfect rotation in Fig. 7.8b is due to the dynamic phase ϕ_{dyn}. The imperfect STIRAP efficiency causes shrinking of the 0-pole of the Bloch sphere. The asymmetry about the z-axis is likely caused by imperfect Ramsey pulses.

This single-qubit phase gate is not the most efficient way to introduce a π-phase to a single qubit; this is rather a demonstration of a quantum gate involving a trapped Rydberg ion. This single-qubit gate may be extended to a two-qubit phase gate in which phases are generated by strong interaction between Rydberg ions. Such a gate—which involves excitation of Rydberg states via STIRAP—was proposed in [20, 21]. This gate requires strong interaction between Rydberg ions. MW-dressed Rydberg states can have large dipole moments [22, 23]. Strong dipole-dipole interaction between such MW-dressed Rydberg states have recently been measured in our laboratory. We are currently trying to implement the two-qubit phase gate.

References

1. Fleischhauer M, Imamoglu A, Marangos JP (2005) Electromagnetically induced transparency: optics in coherent media. Rev Mod Phys 77:633–673
2. Higgins G, Pokorny F, Zhang C, Bodart Q, Hennrich M (2017) Coherent control of a single trapped Rydberg ion. Phys Rev Lett 119:220501
3. Higgins G, Pokorny F, Zhang C, Hennrich M (2019) Highly-polarizable ion in a Paul trap. arXiv:1904.08099
4. Sillanpää MA et al (2009) Autler-Townes effect in a superconducting three-level system. Phys Rev Lett 103:193601
5. Foot C (2005) Atomic physics. Oxford University Press, Oxford
6. Johansson J, Nation P, Nori F (2013) QuTiP 2: a Python framework for the dynamics of open quantum systems. Comput Phys Commun 184:1234–1240
7. Browaeys A, Barredo D, Lahaye T (2016) Experimental investigations of dipole–dipole interactions between a few Rydberg atoms. J Phys B 49:152001
8. Vitanov NV, Rangelov AA, Shore BW, Bergmann K (2017) Stimulated Raman adiabatic passage in physics, chemistry, and beyond. Rev Mod Phys 89:015006
9. Bergmann K, Vitanov NV, Shore BW (2015) Perspective: stimulated Raman adiabatic passage: the status after 25 years. J Chem Phys 142:170901
10. Cubel T et al (2005) Coherent population transfer of ground-state atoms into Rydberg states. Phys Rev A 72:023405
11. Deiglmayr J et al (2006) Coherent excitation of Rydberg atoms in an ultracold gas. Opt Commun 264:293–298

12. Sparkes BM et al (2016) Stimulated Raman adiabatic passage for improved performance of a cold-atom electron and ion source. Phys Rev A 94:023404
13. Saffman M, Walker TG, Mølmer K (2010) Quantum information with Rydberg atoms. Rev Mod Phys 82:2313–2363
14. Berry MV (1984) Quantal phase factors accompanying adiabatic changes. Proc R Soc A 392:45–57
15. Yale CG et al (2016) Optical manipulation of the Berry phase in a solid-state spin qubit. Nat Photonics 10:184–189
16. Zanardi P, Rasetti M (1999) Holonomic quantum computation. Phys Lett A 264:94–99
17. Duan L-M, Cirac JI, Zoller P (2001) Geometric manipulation of trapped ions for quantum computation. Science 292:1695–1697
18. Chuang IL, Nielsen MA (1997) Prescription for experimental determination of the dynamics of a quantum black box. J Mod Opt 44:2455–2467
19. James DFV, Kwiat PG, Munro WJ, White AG (2001) Measurement of qubits. Phys Rev A 64:052312
20. Møller D, Madsen LB, Mølmer K (2007) Geometric phase gates based on stimulated Raman adiabatic passage in tripod systems. Phys Rev A 75:062302
21. Rao DDB, Mølmer K (2014) Robust Rydberg-interaction gates with adiabatic passage. Phys Rev A 89:030301
22. Müller M, Liang L, Lesanovsky I, Zoller P (2008) Trapped Rydberg ions: from spin chains to fast quantum gates. New J Phys 10:093009
23. Li W, Lesanovsky I (2014) Entangling quantum gate in trapped ions via Rydberg blockade. Appl Phys B 114:37–44

Chapter 8
Summary and Outlook

The theory investigation by Müller et al. [1] predicted that trapping of Rydberg ions in a linear Paul trap is feasible and that strong interactions between trapped Rydberg ions may be used for carrying out fast quantum gates. Trapped Rydberg ions are thus an interesting platform for quantum information processing, and various other proposals are summarised in Sect. 1.3.1. These proposals have motivated two experimental investigations of trapped Rydberg ions: $^{40}\mathrm{Ca}^{+}$ ions are excited to Rydberg states using a single-photon excitation scheme in Mainz, while $^{88}\mathrm{Sr}^{+}$ ions are excited to Rydberg states by two-photon transitions in the experiment described in this dissertation.

The two-photon excitation scheme makes it easier to work within the Lamb–Dicke regime, as discussed in Chap. 4. This allows us to attain narrower Rydberg resonance lines than the Mainz experiment. With narrow resonance lines we have investigated effects of the strong trapping electric fields on the sensitive Rydberg ions, which are presented in Chap. 6. After trap effects were mitigated, Rydberg states were coherently excited and a single-qubit phase gate was carried out, as described in Chap. 7.

The next step towards using trapped Rydberg ions for quantum information processing is to carry out two-qubit gates which take advantage of strong Rydberg interactions. This is the Ph.D. thesis work of Fabian Pokorny. Towards this goal Fabian and Chi Zhang have coupled Rydberg $S_{1/2}$- and $P_{1/2}$-states using microwave (MW) radiation. The MW-dressed states have dipole moments and ions in these states interact strongly with each other by dipole-dipole interaction. When Rabi oscillations are driven between the low-lying state $|0\rangle$ and a MW-dressed Rydberg state Fabian and Chi have observed a partial Rydberg blockade—this is a signature of interacting Rydberg ions.

Different two-qubit gates may be implemented using strong Rydberg interactions. A Rydberg blockade gate can be carried out if the Rydberg-excitation Rabi frequency is much lower than the interaction strength, as was done with neutral Rydberg atoms [2]. The phase gate demonstrated in Sect. 7.4.3 may be extended to a two-qubit

G. Higgins, *A Single Trapped Rydberg Ion*, Springer Theses,
https://doi.org/10.1007/978-3-030-33770-4_8

controlled phase gate [3, 4] if a Rydberg-excitation Rabi frequency much higher than the interaction strength is used.

After typically several hundred excitations to Rydberg states ions are lost by double ionisation. Until recently these events interrupted the experiment by around 20 min and made it cumbersome to collect data. Ablation loading of $^{88}Sr^{+}$ ions has been introduced to the experiment by Andreas Pöschl and Quentin Bodart, and the ion loading time is reduced; double ionisation events now interrupt the experiment by less than 30 s. Future Rydberg ion experiments may find it invaluable to shuttle trapped ions between a loading zone containing an ion reservoir and an experiment zone, as was envisaged in a Mainz design [5].

References

1. Müller M, Liang L, Lesanovsky I, Zoller P (2008) Trapped Rydberg ions: from spin chains to fast quantum gates. New J Phys 10:093009
2. Isenhower L et al (2010) Demonstration of a neutral atom controlled-NOT quantum gate. Phys Rev Lett 104:010503
3. Müller D, Madsen LB, Mølmer K (2007) Geometric phase gates based on stimulated Raman adiabatic passage in tripod systems. Phys Rev A 75:062302
4. Rao DDB, Mølmer K (2014) Robust Rydberg-interaction gates with adiabatic passage. Phys Rev A 89:030301
5. Feldker T (2017) Rydberg excitation of trapped ions. PhD thesis, Johannes Gutenberg-Universität Mainz

Printed in the United States
By Bookmasters